201 Solar Eclipses
with the Transit of Venus

A guide for those who want to maximise their experience of eclipses
by Sheridan Williams FRAS

AUTHOR

Sheridan Williams is Director of the Computing Section of the British Astronomical Association, Fellow of the Royal Astronomical Society and Secretary of the Open University Astronomy Club. In 1966 he built his own telescope and has since travelled to, and seen, 12 total and two annular eclipses, and the 2004 Venus transit. His publications include a book on UK solar eclipses, and Bradt's *Total Solar Eclipse 2006,* and *2008 & 2009.* He was also consultant for *Africa and Madagascar: Total Eclipse 2001 & 2002* published by Bradt, and his articles have appeared in *Astronomy Now* magazine, and eclipse papers in the *BAA Journal.* Sheridan has appeared on television and radio, and presented Sky TV's total eclipse programme from Cornwall in 1999. He writes regular features for his local newspaper and lectures widely to astronomical societies and other interest groups. He is also an astronomy travel consultant and guide, leading groups to see various astronomical events such as aurorae and meteor showers. He is a guide at the National Museum of Computing at Bletchley Park whose star attraction is Colossus – the world's first programmable digital computer.

First published January 2012

Bradt Travel Guides Ltd, IDC House, The Vale, Chalfont St Peter, Bucks SL9 9RZ, England
www.bradtguides.com
Published in the USA by The Globe Pequot Press Inc, PO Box 480, Guilford, Connecticut 06437-0480

Text copyright © 2012 Sheridan Williams
Photographs copyright © Individual photographers (see below)
Some maps © 2012 Bradt Travel Guides
Project manager: Maisie Fitzpatrick

The author and publishers have made every effort to ensure the accuracy of the information in this book at the time of going to press. However, they cannot accept any responsibility for any loss, injury or inconvenience resulting from the use of information contained in this guide. All rights reserved. No part of this publication may be reproduced, stored in a retrieval system, or transmitted in any form or by any means, electronic, mechanical, photocopying, recording or otherwise without the prior consent of the publishers. Requests for permission should be addressed to Bradt Travel Guides Ltd in the UK or to the Globe Pequot Press Inc in North and South America.

ISBN: 978 1 84162 366 5

British Library Cataloguing in Publication Data
A catalogue record for this book is available from the British Library

Photographers Paul Coleman (PC); Dreamstime: Olga Khoroshunova (OK/DT); Mike Foulkes (MF); Glenn Schneider (GS); Sheridan Williams (SW); Wikipedia Commons: Malcolm Jacobson (MJ/WC), Tourism NT (TNT/WC); Ariadne Van Zandbergen (AZ)
Front cover Diamond ring, Morombe, Madagascar, June 2001 (MF)
Back cover Total eclipse from Antalya, Turkey (PC)
Title page Totality (SW)

Other credits
Many of the eclipse maps and data are courtesy of Fred Espenak (NASA), Jay Anderson, Michael Zeiler and Xavier Jubier

Typeset from the author's disk by Wakewing, High Wycombe
Production managed by Jellyfish Print Solutions; printed in the UK

Contents

	Introduction	IV
Chapter 1	**Total Solar Eclipses** Explanations 1, Eclipses in history 10	1
Chapter 2	**The Total Solar Eclipse of 13/14 November 2012** Where it goes 12, The climatology of the 2012 eclipse track 13, Top end and far north travel 16, Summary 16	12
Chapter 3	**The Hybrid Solar Eclipse of 3 November 2013** Hybrid eclipses 19, Where it goes 20, The climatology of the 2013 eclipse track 21	19
Chapter 4	**Annular Eclipses** Introduction 26, The annular solar eclipse 20/21 May 2012 28, The annular solar eclipse of 10 May 2013 30	26
Chapter 5	**The Venus Transit of 6 June 2012** Introduction 33, Where it can be seen 33	33
Chapter 6	**Planning, Preparation and Photography** Health and safety 36, What to take 38, Photography 39	36
Chapter 7	**Conclusion**	46
Chapter 8	**Further Information** Books 48, DVDs and videos 49, Internet 49, Maps 50	48

Introduction

> There is nothing in Nature to rival the glory of a total eclipse of the Sun. No written description, no photograph, can do it justice.
>
> Sir Patrick Moore OBE, veteran British astronomer, 1999

A total solar eclipse is the most moving experience on Earth for many of those who witness one. While some feel unrestrained joy at the sight of totality, others feel an equally powerful sense of desolation, such as an observer in England in 1927:

> It was as if the whole Earth were smitten with a mortal sickness … It was all inexpressibly sad and utterly desolate, and when the small golden crescent appeared again behind the Moon, I involuntarily uttered the words 'Thank God for the Sun'.

No-one can really explain why a total eclipse of the Sun has the power to unleash such varied emotions, leaving us so momentously disorientated. After all, the movements of the Earth, Moon and Sun have been predictable for thousands of years – and explicable for hundreds. They are hardly a surprise anymore. Yet they still overwhelm us. I believe it is because of the instinctive realisation, as the eclipse unfolds, that nothing whatever can prevent it. We already know this in a rational way, of course, but the eclipse drives it into our understanding with unequalled drama. Another observer in 1927 writes:

> It is well that we should now and then be made to realise the might and majesty of the universe in which we are the greatest and the least.

I thought I was well briefed when I travelled to Alderney in the English Channel to view the total solar eclipse of August 1999. As the *Daily Telegraph*'s science correspondent I had written thousands of words about eclipses in the preceding weeks, although I had never actually seen one.

But I discovered that I was not prepared at all. My response to the Moon gliding in front of the Sun, as I watched with jubilant crowds from atop the island's crumbling fort, was totally unexpected: it was fear. Suddenly, powerfully, I did not want the eclipse to happen. But it did, of course, and gradually my fear disappeared. It was replaced by exhilaration, echoed and intensified by the screams and cries of my fellow viewers.

Wherever you go, you will be joining a distinguished club of 'eclipse chasers' whose history is at least 200 years old and arguably dates back as far as Captain Cook. Their numbers are exploding as word spreads about this natural wonder.

Before you go, however, read this guide. It will help you choose your setting, take memorable images, and protect your eyes.

We – the author and editors – have done everything we can to make your eclipse experience a splendid one. But there is one thing we cannot do – as I know too well – and that is to prepare you for your own reactions.

Aisling Irwin

Total Solar Eclipses

Every time the Moon passes between the Earth and the Sun, casting its shadow on our planet, the experience for the humans who watch it is different from the last. Every solar eclipse is governed by a multitude of rhythms and subtleties that guarantee its individuality. The Sun's activity, its flares and prominences, the exact positions of the Moon and the Earth – all these factors conspire with numerous others to ensure that each eclipse is unique. Similarly, the aesthetics of an eclipse are influenced by the time of day, the region of the world, the weather and, to a surprising extent, the surrounding crowds or emptiness.

Yet, despite this capacity for infinite variation, eclipses are governed by the enduring laws of celestial mechanics that guarantee us totalities for hundreds of millions of years to come.

EXPLANATIONS

WHY ECLIPSES HAPPEN A total solar eclipse occurs when the Moon moves between the Sun and the Earth, fully blocking our view of the Sun.

This happens because of a heavenly coincidence: the Sun and the Moon, when viewed from the Earth, appear to be the same size. The Sun may seem at first to be the larger but this is because it is so bright. If you stick your thumb and forefinger out to gauge their apparent (or angular) sizes you will find they are the same.

In reality the Sun's diameter is 400 times that of the Moon. But, with perfect compensation, it sits 400 times further away. This symmetry has not always existed but scientists have calculated that things will remain this way for another 650 million years, after which the Moon will drift too far from the Earth for it ever again to be able to cover the Sun completely.

During the year the Earth takes to complete its journey around the Sun, the

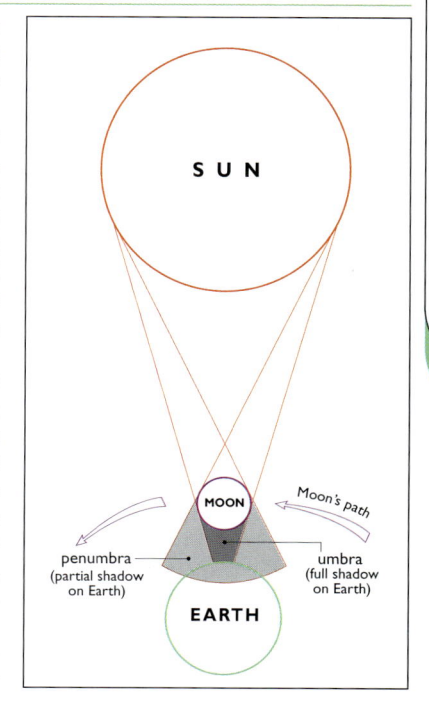

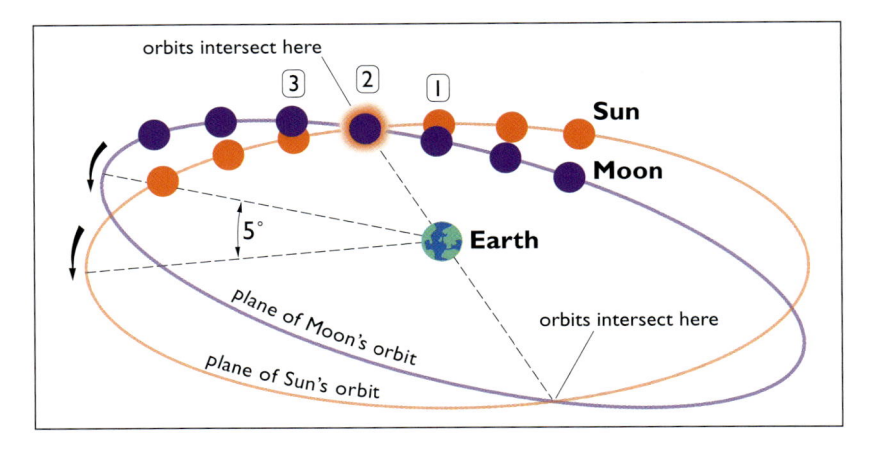

Moon speeds around the Earth about 13 times. Once in each of these cycles (known as synodic months) the Moon passes between the Sun and us and we see a new Moon. We do not see a monthly solar eclipse, though, because the Moon's orbit is at a slight angle to that of the Earth – so the Sun and Moon 'miss' each other as they pass across our skies.

To understand the orbits relevant to an eclipse it is easier to consider a model in which the solar system is inverted, with the Earth at its centre. This model, still in use for some purposes today, represents not the reality of planetary orbits but what we actually see when we look at the skies. In this model, the Sun's apparent path across the sky is known as the ecliptic.

As viewed from the Earth the Moon circles us every 27.5 days, and the eclipse year of 346.62 days is the time for the Sun (as seen from the Earth) to complete one revolution with respect to the same lunar node. The Moon catches up with the Sun every 29.5 days and would pass between it and Earth, causing a solar eclipse, if it weren't for those inclined orbits (see diagram above). But the two orbital paths do, of course, intersect at two points. These points – or nodes – represent two periods each year during which the Moon can indeed pass in front of the Sun. When the Moon passes exactly in front of the Sun we either see a total or annular eclipse; these are called 'central' eclipses because the Moon passes centrally across the Sun.

If the Sun, creeping along its yearly path, happens to be almost exactly at one of the two intersections when the Moon is racing round on its monthly cycle, there will be a central solar eclipse. If the Sun is a little further from the intersection there is still the chance that the Moon will at least overlap with it, creating a partial solar eclipse.

In fact, the Sun moves so slowly that it is impossible for it to get past either intersection without the Moon catching up with it at least once, and therefore guaranteeing a solar eclipse of some sort. Sometimes the Moon manages to travel all the way round and catch up with the Sun before it has fully left the intersection – causing two partial eclipses a month apart. Whatever happens during these eclipse 'seasons', there will always be at least two solar eclipses of some sort each year.

Those are the basics behind a solar eclipse. In practice, however, there are various distortions and complexities that give each eclipse its signature. For example, both the Sun and the Moon appear to follow elliptical paths around the Earth rather than circular ones, with the result that their distances from our planet vary. If a solar eclipse occurs when the Moon is at its furthest from the Earth and the Sun at its closest, a rim of Sun remains around the black disc of the Moon – and we see an annular eclipse.

THE SAROS Eclipses pursue other, more long-term rhythms, the most distinctive of which is the Saros cycle, discovered by the Chaldeans possibly as early as 400BC and used over subsequent millennia to predict eclipses. The Chaldeans found that eclipses follow 18-year cycles, with each new solar eclipse bearing similarities to the one that occurred 18 years previously.

An eclipse casts a shadow in an arc across the Earth. For that shadow to be identical to the shadow cast by another eclipse, the relative orientations of the Earth, Moon and Sun would have to be exactly the same. This can occur only on the rare occasions when the Moon's position in its orbit around the Earth and the Earth's position in its orbit around the Sun are repeated. This does not happen every time. For example, when one eclipse year has passed and the Earth begins its orbit around the Sun again, the Moon is three-quarters of the way through a cycle round Earth; so it is not back in the same position as it was when the Earth last began an eclipse year. But after the passage of approximately 18 years 11.32 days the Moon is just coming to the end of its monthly cycle when the Earth is coming to the end of its yearly cycle – and they pass through positions very similar to those of 18 years previously. Thus, when an eclipse occurs over a certain region of the Earth, we can predict that, 18 years later, an eclipse will occur at a similar latitude. Because this synchronisation is not perfect, the next eclipse will be displaced from its 18-year-old sister by about 11 days and one-third of the world to the west. Successive solar eclipses in a Saros series also gradually shift their tracks northward or southward. Odd-numbered Saros series shift southward, and even-numbered shift northward on the Earth.

The total eclipse of 2012 is the 45th member of 72 eclipses in Saros series 133. All eclipses with an odd number occur at the Moon's ascending node and the Moon moves southward with each member in the family. The series started on 13 July 1219 and includes 190 partials, six annulars, one hybrid and 46 totals. The series ends on 5 September 2499, and thus spans a period of 1,280 years. The longest total eclipse in this Saros was in 1850 and lasted a creditable six minutes and 50 seconds.

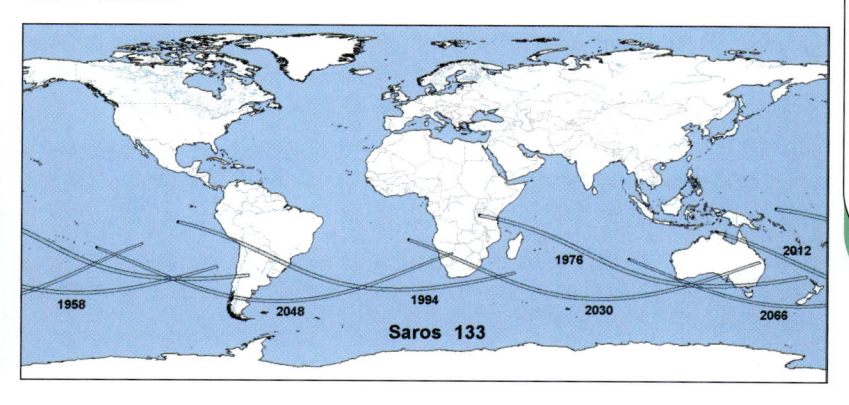

Saros 133

The total eclipse of 2013 is the 23rd member of 72 eclipses in Saros series 143. The series started on 7 March 1617 and includes 30 partials, 26 annulars, four hybrid and 12 totals. The series ends on 23 April 2897, and thus spans a period of 1,280 years. The longest total eclipse in this Saros was in 1887 and lasted three minutes and 50 seconds. The eclipse of 2013 is the longest hybrid eclipse in the series and the only one of type annular to total (A–T).

3

The maps on this and the previous page show the seven eclipses centred on 2012 and 2013 in Saros 133 and 134. Notice that they shift around the globe, moving west by eight hours each time.

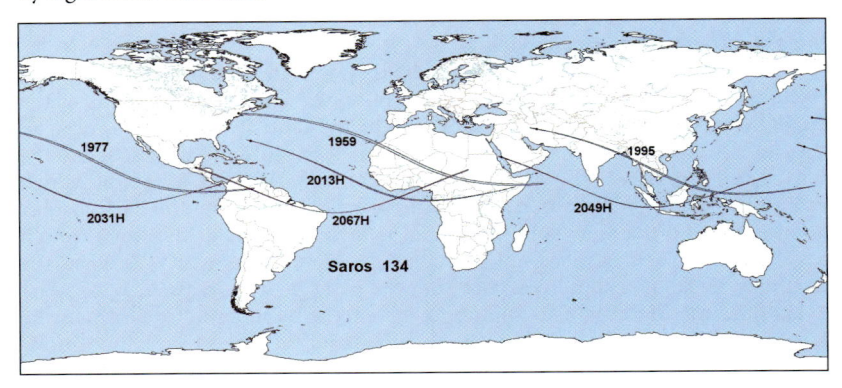

In Saros 134 eclipses appended T are total and those appended H are hybrid.

ANATOMY OF THE SUN Our knowledge of the Sun has grown significantly in recent years, partly because of the spacecraft SOHO, which creates a permanent eclipse between itself and the Sun, allowing constant monitoring of its outer layers.

The core of the Sun is a furnace of burning hydrogen at 15 million °C, hidden from view by a shell of opaque gases 696,000km thick. Working outwards, through layers of radiation and convection, we reach the **photosphere**. This is the Sun's visible surface – a boiling sea of rising and sinking gases of 5,500°C. It is the layer where sunspots emerge and vanish. The photosphere is only about 300km deep.

Above this layer is the **chromosphere**, a vivid orange colour, just 2,500km thick. Its temperature varies strangely, the lower levels being about 4,000°C and the upper layers hotter, at about 10,000°C. The chromosphere is not smooth but covered with sporadic projections of gas which can reach 700km high, as far as the next layer – the corona.

The **corona** is a haze stretching far from the Sun, growing ever thinner yet still detectable as a tenuous haze beyond Earth. Those parts of the chromosphere which stretch into the corona soar in temperature to 1 million °C. This heat accelerates charged particles outwards, imbuing them with enough speed to escape the Sun's gravitational field. The gases rising from the surface of the Sun expand as they go, and so their density decreases and they are more easily warped by the Sun's magnetic fields. This twists them into loops and arches known as **prominences**. Sometimes a magnetic loop breaks and billions of tons of gas shoot into space, creating a coronal mass ejection.

X-ray pictures reveal the Sun's outer layer to be full of roller-coaster loops and exploding gases interspersed with darker, quieter regions known as **coronal holes**. The beauty of a total solar eclipse is that, with the glare of the photosphere obscured, you can for once witness the chromosphere, prominences and corona.

EFFECTS SEEN DURING A TOTAL SOLAR ECLIPSE Although the arrangement of the Sun, Moon and Earth may seem identical, many factors combine to change the look of every eclipse. For example:

- How far the Sun is above the horizon
- The amount of cloud and dust in the sky
- The difference between the sizes of the Sun and Moon
- The libration (oscillating motion) of the Moon
- Where the Sun is in its activity cycle

These factors manifest themselves in the shape and size of the corona, the landscape, shadow bands, and the location of Baily's beads and the diamond ring.

THE CORONA AND CHROMOSPHERE For descriptions of the chromosphere, corona, prominences and the photosphere see *Anatomy of the Sun*, page 4.

During periods of solar minimum, the corona is roughly arranged in the equatorial regions. However, during the Sun's active periods, the corona is more evenly distributed, though it is most prominent in areas with sunspot activity. The solar cycle spans approximately 11 years. The photographs below show solar minimum on the left and solar maximum on the right:

(SW) 2008

(SW) 1991

Baily's beads The Moon does not have a smooth outline, and as the Sun is just about to be covered (second contact) its last vestiges shine through the lunar valleys. Francis Baily (see page 11) was the first to discuss these and they are now known as Baily's Beads. To predict in advance where beads and the diamond ring will occur on the Sun's disc, you need to refer to the 'lunar limb profile' chart. This is different for every eclipse and even changes depending on where in the eclipse track you observe. The **lunar limb profiles** have been provided for the most likely locations to be visited for each eclipse. C2 and C3 mark the points of second and third contact. If you were located near the edge of the eclipse track you would witness extended duration beads at the north and south positions. If the Moon and Sun are exactly the same size you

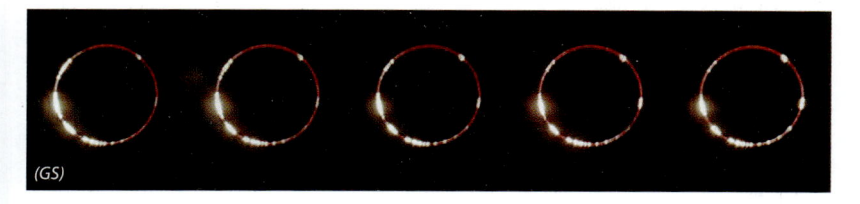

(GS)

get beads all around, as shown on the amazing sequence of photographs taken by Glenn Schneider while flying at 45,000ft in a Cessna Citation over Greenland (see previous page).

The diamond ring When all the beads have disappeared except the last one, this is the diamond ring. It will also appear exactly as totality ends. The photographs below show the beginning and end diamond ring and were taken by me in Libya in 2006.

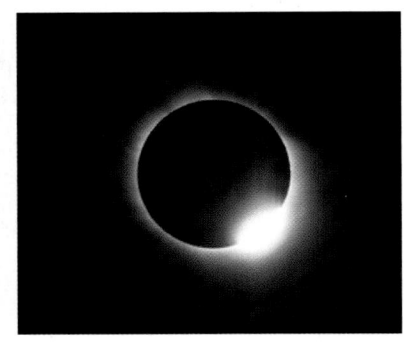

On examining the lunar limb profile (see page 5), you will see huge deep valleys at both C2 and C3. This could produce the largest diamond rings ever seen.

Shadow bands Shadow bands are thin wavy lines of alternating light and dark that can be seen moving and undulating in parallel on plain-coloured surfaces immediately before and after a total solar eclipse. Shadow bands result from the illumination of the atmosphere by the thin solar crescent a minute or so before and after totality.

In 1842, George B Airy, the Astronomer Royal, saw his first total eclipse of the Sun and recalled shadow bands as one of the highlights:

> As the totality approached, a strange fluctuation of light was seen upon the walls and the ground, so striking that in some places children ran after it and tried to catch it with their hands.

WHAT TO WATCH FOR AS THE ECLIPSE UNFOLDS: STEP BY STEP

> An unearthly gloom enveloped us, and the atmosphere grew chill, so that we shivered perhaps as much with awe as with the cold …
>
> R A Marriott, *A British Eclipse*, 1927

Once you are well inside the path of totality (which you can work out using the maps and tables in this booklet) you can settle down to watch the spectacle unfold. Do not worry too much about memorising the precise timings of the stages of the eclipse – as long as you know roughly (to the nearest 15 minutes) when to look, nature will guide you through the experience.

When the Moon first encroaches on the Sun, the moment known as **first contact** has arrived. It is followed roughly one hour later by **second contact** – the moment when the Moon drifts fully in front of the Sun and totality begins. **Third contact** heralds the end of totality because the Moon starts to shift away from the Sun. **Fourth contact** is when the two discs finally part.

At first an eclipse is almost unnoticeable; it's just a shallow scoop from one edge of the Sun which hardly diminishes the daylight. The fading of the day happens very slowly and, until the Sun is over 80% covered, there is little more than a faint dullness. Even when the Sun is 80% covered most of us will notice little change in the light because our brains are so used to compensating for the effects of heavy cloud. But the Moon crawls relentlessly onwards and, when it has covered 90% of the Sun, the light will fail noticeably and the land will turn a bluish grey. With about ten minutes to go, some people will shield one eye to accustom it to darkness, so they will be better able to appreciate the wisps and auras that characterise totality.

From about five minutes before totality you should start watching for brighter planets such as Mercury and Venus near the Sun. During the last few minutes before second contact, daylight will disappear fast. Now is the moment to watch the ground for **shadow bands** rippling across lightly coloured surfaces, as if on a lake. The ripples appear because of a lensing effect caused by the Earth's atmosphere which bends and focuses the light streaming towards Earth from the Sun's thin remaining crescent. You can see them more easily if you lay a white sheet on the ground. Note also tiny crescent Suns projected onto the ground through gaps between the leaves of trees and bushes. You may like to take a piece of card punctured with holes (a

THE STAGES OF A TOTAL SOLAR ECLIPSE

First contact: The moment when the edge of the Moon first encroaches on the Sun.

The Moon approaches 50% coverage of the Sun but the ambient light has changed little.

Second contact: The Diamond Ring: a brilliant gleam of light appears as the last bit of the Sun shines through a lunar valley.

Totality: The eclipse is total. When it ends we have third contact. The diamond ring will appear on the other side.

Third contact: As totality ends, a diamond ring usually appears on the opposite side to the one at second contact.

Fourth contact: The Moon takes its leave of the Sun. The eclipse is over.

The direction of the Moon's travel varies depending on your location. It can traverse from top left to bottom right in Cairns in 2012 and from bottom right to top left in Africa in 2013.

knitting needle will do to make the holes) in case there is no vegetation to do the job for you.

At around this time, and assuming that it is not already a windy day, you will be aware that there is a distinct breeze. This can be sufficient to blow any paperwork around that may be lying about. This was thought to be a coincidence, but can be easily explained because the air inside the Moon's shadow is cooler (and hence of different density) than the air outside the shadow. This causes air movement and hence wind. The wind was very noticeable in the Libyan desert in 2006, and also in the Australian outback in 2002.

About 15 seconds before the Sun is completely eclipsed, you will see specks of light appear around the dark disc of the Moon, just like a string of pearls. These are **Baily's Beads**, the last few rays of sunlight shining through valleys on the edge of the Moon. They will swiftly die away until just one remains – a dazzling jewel known as the 'diamond ring', which shines brilliantly for just three or four seconds and then vanishes.

From the moment of the diamond ring's appearance it is safe to watch the eclipse with the naked eye, though you may pay a penalty for viewing the diamond by being unable at first to distinguish the features of the ensuing **totality**. You will know when totality has arrived. Suddenly the Moon's shadow rushes across the land towards you and your world is plunged into twilight. At this point the emotion can be overwhelming.

Now the secrets of the Sun's outer layers, usually hidden by the glare of its fiery centre, are briefly revealed. For the first few seconds of totality you may see a vibrant, pinkish-red rim at the edge of the Sun. This is the light from the Sun's lower atmosphere, the chromosphere. As the Moon moves onwards it will swiftly cover this too (during very short eclipses the **chromosphere** never really disappears because the Moon is only minutely larger than the Sun).

An effect that should last a little longer will be the appearance of several deep red clouds, like smoke drifting from the surface of the Sun. These are the solar **prominences**, stretching to a distance of up to one-twentieth of the Sun's diameter (see the picture on the title page that shows a lovely prominence during the 2006 eclipse).

At full totality you will see the black disc of the Moon perfectly surrounded by the pearly light of the Sun's corona. Wispy plumes and streams of coronal light will dance outwards.

If you can drag your eyes away from the eclipse itself, you will have an unprecedented opportunity to view the daytime positions of planets and stars which began to emerge just before totality.

Don't forget to check the landscape, where nature will respond as if night has fallen. Flowers will close their petals and disorientated bees will cease their flight. Insects, maintaining their daytime silence until the moment of totality, will suddenly sing as if it were night, and mosquitoes will start biting. Animals who were grazing peacefully may become nervous and confused, while

(SW)

the birds above them will flock into the trees with great fuss. Cows have been known to line up and walk to their shelters.

In the distance, where the eclipse is only partial, the horizon will be an evening pink. The temperature plunge will be noticeable.

If you are near the centre line, the diamond ring emerges again at roughly the opposite edge of the Sun from before. Totality has finished and the sequence will begin again in reverse. At the diamond ring's appearance look away or use your eclipse viewer. At fourth contact, the eclipse is over.

If the day is thick with cloud then most of these sights will be hidden, but partial cloud and thin cloud will not necessarily spoil the experience. The eclipse carries with it its own weather patterns, caused by the wild temperature variations that the moving shadow creates on Earth. This can cause clouds miraculously to disperse during the seconds leading up to totality – but more often it can cause them swiftly to materialise out of a previously clear sky (rapid cooling of the air can have either effect, depending on the humidity and pressure). Trying to catch a glimpse of totality through the spaces between thick clouds has its own thrills – even one fleeting view seems a tremendous achievement and is deeply exciting. If there is broken cloud you may see patterning and colours in the sky. If there is thin cloud, the corona may still be visible. Finally, for those besieged by the weather, watch out for a shadow that the Moon may cast onto the cloud above you.

CHOOSING WHERE TO SEE THE ECLIPSE Many people consider that the only place to go to see an eclipse is on the centre line as this maximises the duration of totality. However, there are many things to consider when choosing your location:

Weather Simple: choose the least cloudy location. Be aware also of the possibility of pollution, and smoke from forest fires and local chimneys. Sandstorms can occur in some areas too. Do not choose a location near to roads where car headlights can spoil the picture. Beware of hotel floodlights that may come on as darkness descends.

Wildlife Animals react to large eclipses and you may want to witness their behaviour.

Scenery Deserts have a stark beauty, and are a natural choice for low cloud cover. But foliage gives special effects by projecting the Sun through its leaves. How about photographing the eclipse through a rock arch, or with a tree in the foreground?

Sun's altitude I've heard people say that the higher the eclipse in the sky, the longer the duration and the least chance there will be clouds in your line of sight. This is not necessarily true. Consider also that some of the most dramatic eclipses occur at sunrise and sunset, as was shown in 2001 in Madagascar, 2002 in Australia and 2003 in Antarctica.

Eclipse duration Maximum duration is not necessarily on the centre line. The Moon's mountainous rim can make maximum totality occur slightly away from the centre line. Also consider that in eclipses with a duration of around two minutes or more, you can be 20km from the centre line and still only lose about five seconds of totality.

Eclipse effects There are people who choose to observe totality near the northern or southern limits. They sacrifice the length of totality to see extended Baily's beads and prominences.

ECLIPSES IN HISTORY
HUMANS AND ECLIPSES

> The Sun ...
> In dim eclipse disastrous twilight sheds
> On half the nations, and with fear of change
> Perplexes monarchs.
>
> John Milton, *Paradise Lost*, Book I, lines 587–600 (1667)

Milton's sentiment remains true today. A total solar eclipse bypasses the brain and its scientific understanding, and speaks primitively to the heart. It reminds us of our cosmic insignificance.

Eclipses caused far more perturbation when they were unpredictable – and historians have studied ancient responses to them.

One of the oldest known eclipse tales comes from China and relates to two astronomers, Hsi and Hoe, who became drunk and failed to respond adequately to a total solar eclipse, a crime which was punished by death. If this were true it would mean that the first recorded eclipse occurred between 2159 and 1948 BC. But it is almost certainly a myth, perhaps a morality tale aimed at civil servants of the time.

The first person to predict totality with success was thought for a long time to have been Thales of Miletus. He is said to have foretold the year and position of an influential eclipse that plunged a battlefield into darkness during an encounter between the Lydians and the Medes. Chastened by the experience, the two sides laid down their arms and agreed a peace deal. Today's astronomers cannot see how Thales could have predicted an eclipse given the limited knowledge at the time. It is generally agreed that, while the solar drama may well have ended a war, probably

The 1927 total solar eclipse was visible in North Wales and Lancashire (SW)

on 28 May 585BC, it was not actually predicted.

Ancient civilisations such as the Egyptians, the Chinese and the Babylonians all developed methods for tracking the motions of the Sun and Moon, and had thorough constellation maps. The best-known eclipse archive was developed by the Chaldeans who began to keep precise historical records from 750BC to AD75.

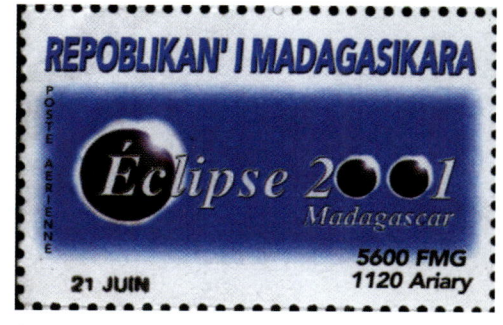

A stamp commemorates the 2001 total solar eclipse over Madagascar (SW)

They developed mathematical theories from these records and, in the 4th century BC (the Hellenistic period), they made the breakthrough of examining past records to look for repeating cycles: eclipse prediction had begun.

Even today, those Babylonian records, stored on thousands of clay tablets, are useful. In 1997 F Richard Stephenson, an astronomer from Durham University in Britain, used a discrepancy between early eclipse records and today's calculations to show that the Earth's rotation is slowing by 2.3 milliseconds per century. His work was published in *Historical Eclipses and Earth's Rotation* (see *Further Information*, page 48).

Scientific study of the Sun during eclipses began in the early 19th century with Francis Baily, who gave his name to the 'string of bright beads' that fringes the Moon just before totality. In 1715, Edmund Halley (of comet fame) produced a paper for the general public encouraging them to look at the May eclipse that crossed London and large parts of Britain. This was Britain's last really good total eclipse and it lasted over four minutes in a cloudless May sky. Eclipses were studied intensely because they allowed astronomers to determine the distance of the Sun and Moon with greater accuracy. An early 20th-century eclipse even helped prove Einstein's theory of relativity.

After these successful public relations activities, scientists began their modern pursuit of eclipse chasing, pointing all manner of instruments at the Sun to exploit the rare moment when its faint corona could be viewed unimpeded.

RECURRING ECLIPSE LOCATIONS

On 10 May 2013 an annular eclipse will occur just north of Cairns, and in 2028 Sydney will witness a total solar eclipse (for details of the 2012 eclipse, see page 12). In 2077 Cairns will witness a 'near miss' total eclipse. As for Gabon, in 1919 the famous eclipse which helped prove Einstein's theory of relativity passed through here, and then in 1987 a hybrid eclipse occurred of virtually zero seconds duration. Given below are the dates of the previous and next total solar eclipses to pass over the locations listed.

Location	Previous	Future
Cairns 2012	1847	2237
Gabon 2013	1987	2165
Kenya 2013	1898	2165

2

The Total Solar Eclipse of 13/14 November 2012

WHERE IT GOES

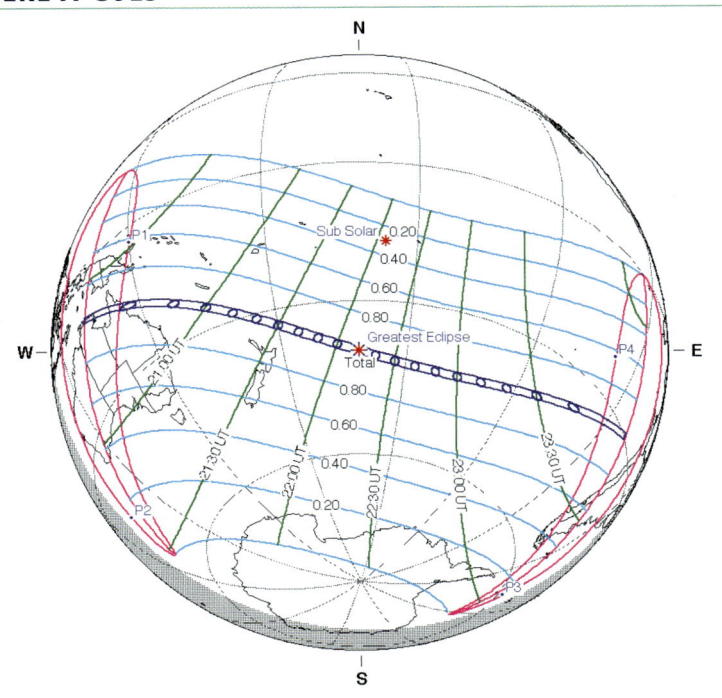

THE ECLIPSE PATH The path of the Moon's umbral shadow starts in the extreme north of Australia (Arnhem Land) at dawn on 14 November and crosses the Gulf of Carpentaria before entering the northern part of Queensland at Edward River. After crossing the Great Dividing Range the centre line passes close to Port Douglas where totality lasts two minutes and three seconds, with Cairns well within the track of totality lasting two minutes. At these locations the Sun has risen to an altitude of 14°, and first contact is ten minutes after sunrise. Off the coast, the centre line passes over Flinder's Reef; it then crosses the south Pacific Ocean without ever touching land again. The eclipse's maximum duration is in the South Pacific and is four minutes and two seconds.

For a detailed map of the track see http://xjubier.free.fr/en/site_pages/solar_eclipses/TSE_2012_GoogleMapFull.html.

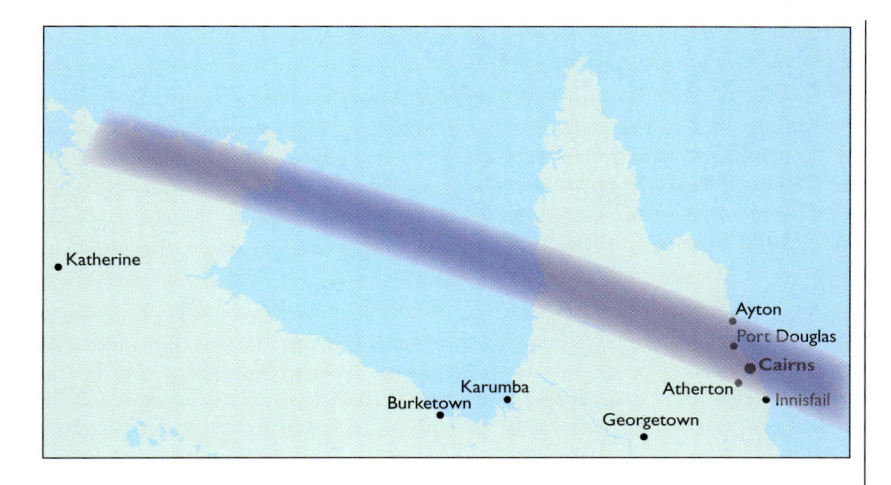

THE CLIMATOLOGY OF THE 2012 ECLIPSE TRACK

OVERVIEW The wettest weather is around Darwin, with progressively less rainfall and fewer days with rain eastward along the track. Cloudiness over the 'Top End' is noticeably higher than over Queensland, but neither has a completely sunny disposition. Northern Territory sites are generally cloudier than those over Queensland. Cairns has the greatest number of sunny days, but inland sites such as Palmerville and Mareeba have fewer cloudy days. As this is an early morning eclipse, with luck it may dawn completely clear, although this is very unlikely. Even a cyclone is possible, though it would be history-making. Most likely is that the day will dawn partly sunny with patches of low-level cloud over the water and along the shoreline – a typical Cairns morning in November, and one guaranteed to stress eclipse watchers as they try to determine the motion of clouds and their chances at a different spot, perhaps in a larger patch of sunshine a few kilometres distant.

Beachfront sites have the advantage of a clear view across the water without concerns that telephone wires or hills might block the view of the rising Sun and early views of this low-altitude eclipse. From central Cairns, a line of hills towards the southeast (Cape Grafton) won't block the view towards the rising Sun, but they are prone to hosting clouds. Instead, try the northern suburbs of the city and the plains to the north as far as Clifton Beach, which offer a gently sloping unobstructed view to the eclipse, with the added advantage of a longer duration of totality. The beaches north of Cairns will likely be crowded with eclipse watchers. Beyond Clifton Beach, the coast rises sharply from the water's edge, opening up again as the highway reaches Port Douglas. There are limited opportunities to stop and set up in the area. Port Douglas lies near the southern edge of a flat coastal plain that extends towards Daintree, but is bounded to the west by the highest part of the Great Dividing Range. The slopes of the mountains are quick to form cloud, but the plain resists that tendency until slightly later in the day. Hotels along the coast may prove superb sites to see the eclipse, but not all of them are well exposed to the ocean and may require a trip down to the beach for the best views. A huge concern for movement in and out of Port Douglas is the Solar Eclipse Marathon that is scheduled to begin at third contact. The cloud and sunshine statistics for Cairns and Port Douglas offer a promising 66% frequency of sunshine in November. Because the sunshine data refers to the entire day, and because the eclipse comes in

> **2012 TOTAL ECLIPSE FACTS**
>
> **Time of greatest eclipse** 22:11:48 Universal Time (UT)
> **Max duration of totality** 4m 02s at latitude 39°57'S, longitude 161° 20'W
> **Altitude of Sun at greatest eclipse** 68°
> **Magnitude** 1.0500; **gamma** −0.372
> **Maximum width of path of totality** 179km
> **Saros series** 133, member 45 of 72

the sunnier morning hours, the probability of seeing it is likely a little greater than the statistics imply. You should set up your equipment up at the top of the beach as the high tides could cover your equipment as the eclipse progresses.

Perhaps the best chance of success along the coast is to 'hunt' from a vehicle moving on the highway north of Port Douglas, where fields are open and places to stop are readily available; provided, of course, that traffic is flowing and there are not hundreds of others with the same idea. This is a viable option only if you start early, in the dark, before you can actually see what the cloud cover looks like – you are gambling on the advantage of mobility versus playing the probabilities on eclipse day by remaining fixed. If, at sunrise, it turns out that the other side of Port Douglas looks better, you may have some difficulty crossing town due to the marathon crowds. All in all, coast observers might be better advised to sit and take whatever nature offers, but it will be a heartbreaker if the eclipse is visible a few hundred metres away yet obscured at your chosen location.

From Port Douglas to Mount Molloy there is a high probability of cloudiness as the air flows through the gap in the mountains. Once at the Mount Molloy crossroads, the ground smoothes out a bit, but trees line the highway and it is difficult to find a suitable view. The surrounding terrain is still very rough, but even on cloudy days it is surprising how small hollows in the topography can chew holes

Your best chances of seeing the 2012 total solar eclipse are likely to be around Port Douglas (MJ/WC)

in the cloud cover, as long as it is not too thick. Sites at all inland locations will have to give up a view of first contact, and there is a possibility that the eclipse itself could be hidden behind a hill if your position is not chosen carefully.

Perhaps the most promising option is to move well behind the high peaks of the Great Dividing Range, to a location near Mount Carbine or some other place where the local terrain provides some shelter from the clouds. There is a danger here that the peaks, which include the highest along the chain, will form cloud on their summits, but one of the peculiarities of the trade-wind zone of the Earth is that high-level winds blow in the opposite direction to those at low level.

The highway from Mount Molloy to Mount Carbine (the Peninsula Development Road) lies in a shallow valley, with the highest peaks to the west and lower ones to the east. At Mount Carbine, Highway 81 turns westward, through a gap in the hills. From a strictly topographic point of view, this might be one of the best conveniently accessible spots along the track. There are stories that some landowners in the area are setting up eclipse-viewing sites for travellers.

After Mount Carbine, the Mulligan Highway (the Peninsula Development Road on Google Maps) heads west and runs parallel to the centre line. After a time, it turns northward to cross the line a little south of Maitland Downs. The town is perilously close to the highest peaks of the range, but at least lies on the leeward side where theory predicts an improvement in cloudiness. The road to Maitland Downs lies within the track of the shadow, so that if an expedition is held up by bad weather, the possibility of seeing the eclipse is not lost. Students from the Atherton State High School scouted the area and their highest recommend locations were 6km and 9km north of the centre line where open fields are present, and one location 6km south of the centre line, at Whumbal Creek. These locations are easily found on Google Earth, though the view there does not suggest anything unusual about the sites.

For the adventurous (in small groups), Palmerville Station, well into the Outback, has the lowest cloud amount (but not the highest number of clear days). The calculated cloud amount for the station, 35%, is much lower than at any other site along the track and is somewhat at odds with other nearby stations. The road to Palmerville Station requires a significant deviation from the shadow path before turning back into it. Travel in the Queensland Outback is not for the novice.

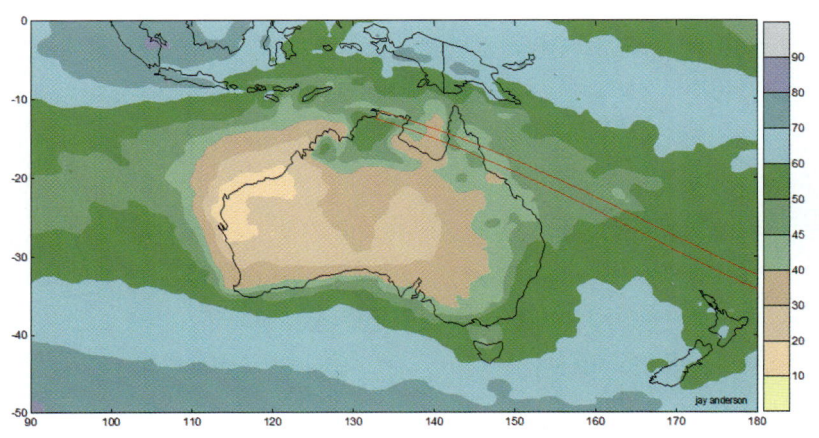

Average daytime cloud cover in November

Climatology report courtesy of Jay Anderson.

TOP END AND FAR NORTH TRAVEL

In the Northern Territory, nearly all the eclipse track is confined to Arnhem Aboriginal Land, and individual tourists are not permitted. However, there will almost certainly be expeditions into the area. After the weather, one problem remains – inland areas are sparsely populated, poorly supplied by roads, subject to flooding and closure, and not capable of handling large groups of tourists.

Travel near Cairns, out onto the Atherton Tablelands, is comfortable, and the central line can be approached along the Peninsula Developmental Road, past Mareeba and Mount Molloy. This will place you behind the higher parts of the Great Dividing Range, on the west side of highly recommended Daintree National Park.

For the more adventurous, who wish to go well inland, the route from Cairns towards Kowanyama is by way of the Burke Developmental Road (BDR). The road is sealed almost to Chillagoe and is comfortably passable for buses and regular vehicles. Beyond this, however, the route roughens and 4x4 vehicles are recommended at all times, especially as November heralds the start of the wet season. The road is sprinkled with mine-workings, old and new, and described as 'good' in the dry. From Chillagoe, the BDR heads northwestward, gradually approaching the southern limit of the shadow zone. About 100km along, and 35km before Gamboola, the road reaches the Mount Mulgrave/Palmerville Road, which turns northward towards the shadow path, and eventually, at Palmerville, reaches the centre line. This is not a route for the uninitiated. Roads can be closed at any time due to flooding and service stations are non-existent. It would be sensible to go with local guides.

SUMMARY

For most of us, this eclipse belongs to Far North Queensland. Cloud prospects are good along the coast, especially in the region between Cairns and Port Douglas, and even better in the Outback on the other side of the Great Dividing Range. Arnhem Land is more remote and cloudier. Vessels offshore would seem to have about the same prospects as the beaches, though their mobility will bring considerable advantage if they have time and room to move.

Travel in the Arnhem Land region is hard-going and only possible for those on organised expeditions (TNT/WC)

LOCAL ECLIPSE CIRCUMSTANCES

Location	Lat (°S)	Long (°E)	1st	2nd	3rd	4th	Alt	Dur
Cairns	16.860	145.780	05:44:42	06:38:30	06:40:33	07:40:18	14°	2m 00s
Port Douglas	16.552	145.506	05:44:30	06:38:08	06:40:13	07:39:47	14°	2m 03s
Peninsula Road	16.486	144.902	05:44:36	06:38:02	06:40:04	07:39:21	13°	1m 58s

Times above are local times (based on Universal Time (GMT) +10 hours); the date will therefore be 14 November. Sunrise for the above locations is 05:34 local time. (Durations are corrected for the lunar limb profile.)

Predictions courtesy of Xavier Jubier.

The lunar limb profile for Cairns 2012

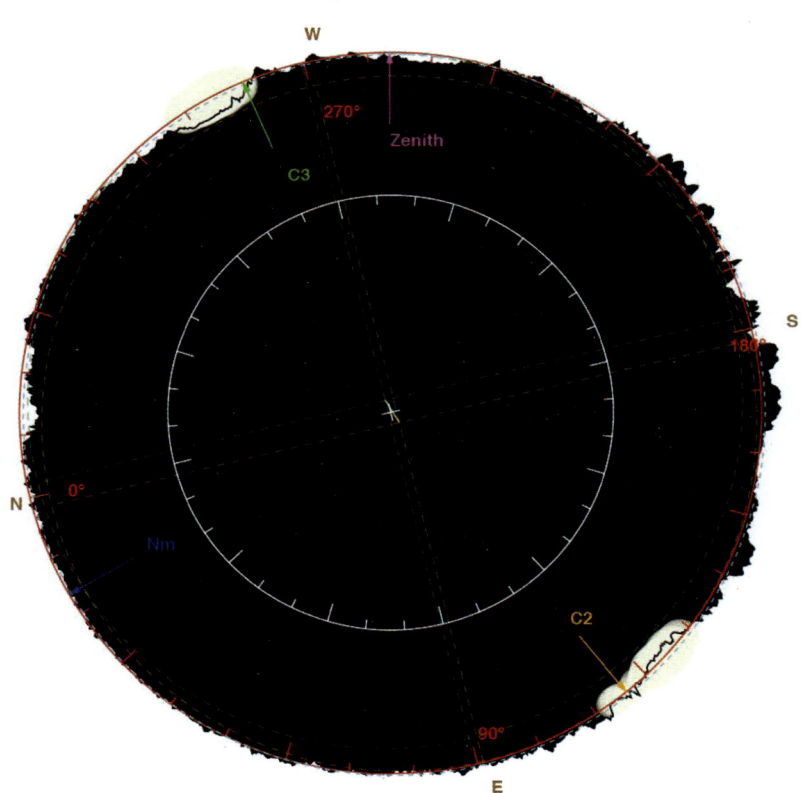

Yellow areas show the expected position of the diamond ring at second and third contact. First contact is at approximately where C3 is marked.

Profile courtesy of Xavier Jubier.

SKY CHART FOR 2012 During totality, planets and brighter stars should be visible. The chart below shows what to look out for.

The Sun is very close to the horizon which may prevent Mercury being visible below it. However, Venus should be very striking, many minutes before totality. Saturn lies between Venus and the eclipse.

Stars that should be visible include Spica (close to Venus), Alpha Centauri, Arcturus, Acrux Regulus (almost overhead); and, on the other side of the sky to the eclipse, Canopus, Sirius, Procyon, Pollux, Betelgeuse and Rigel.

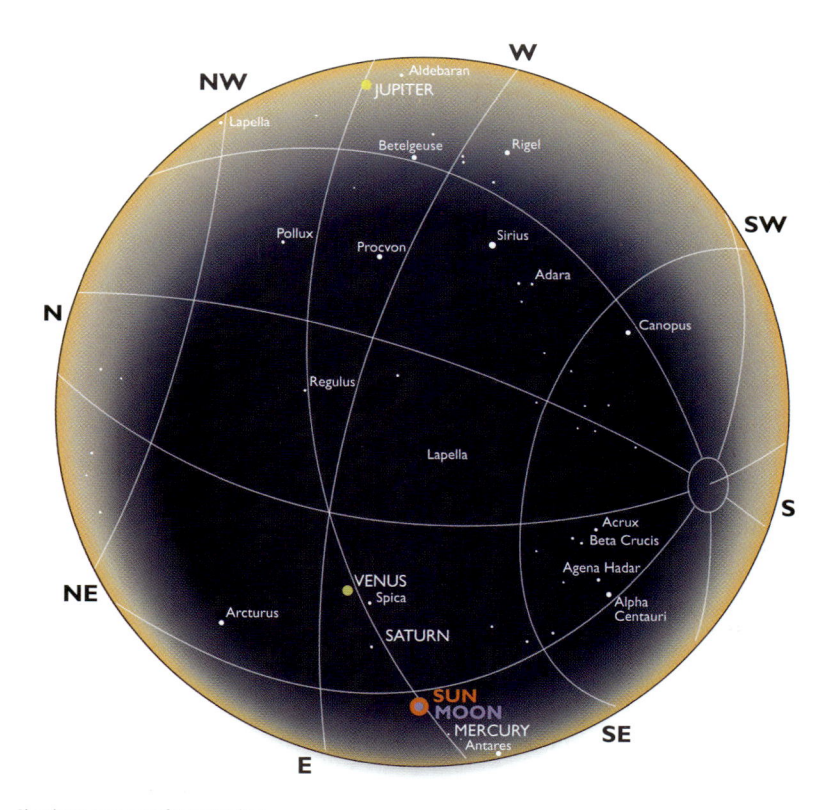

Sky chart courtesy of Xavier Jubier.

3

The Hybrid Solar Eclipse of 3 November 2013

HYBRID ECLIPSES

Annular total eclipses are also known as hybrid eclipses, because they are neither completely annular nor completely total along their whole path. Usually such eclipses begin as annular, become total, then revert to annular again. This is because the Moon is precisely at a point in its orbit where its shadow is only just long enough to touch the Earth's surface. At the beginning of the eclipse the Earth's limb is too far away for the shadow to reach, but as the shadow tracks across the globe it makes umbral contact where the Earth is close enough to the Moon. The diagram below attempts to clarify this.

The 2013 eclipse is a particularly rare type of hybrid eclipse because normally they start annular, go total and then end annular (A–T–A). The most recent of these was on 8 April 2005 where the total phase occurred in the mid-Pacific. Sometimes, however, hybrid eclipses start annular, go total and remain total to the end (A–T), just as the November 2013 eclipse does.

Equally rare are eclipses that start total, go annular and remain so to the end (T–A). The next such eclipse will be on 29 April 2386 and visible across Angola, Zambia, Mozambique and Madagascar.

The rare A–T hybrid eclipses can be explained because the Moon is in such a place in its orbit that it is rapidly approaching the Earth as the eclipse progresses. This closeness prevents the eclipse reverting to annularity. For T–A hybrids, the reverse applies.

HYBRID ECLIPSES

These can happen on rare occasions when the Moon's umbra is not long enough to touch the Earth's surface at the beginning **A** (and **C**) of an eclipse, but is just long enough to strike the surface **B** in the middle of the eclipse.

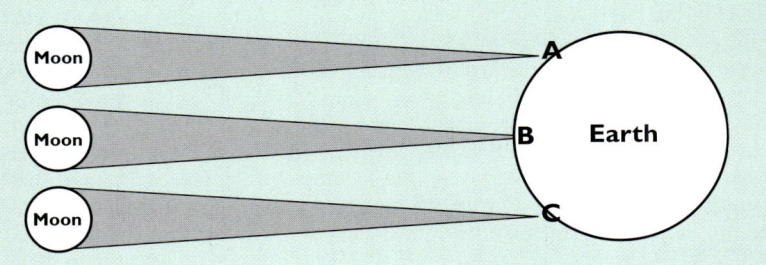

A strange fact is that although there are usually six or seven A–T eclipses in a century, from 2386 until 2500 none occurs for 114 years!

The previous and next hybrid eclipses in Britain both have their total phases over the country: 13 September 1699 at 09:12, duration 20 seconds, type A–T–A; 12 April 2545 at 17:26, duration 11 seconds, type T–A (rare).

WHERE IT GOES

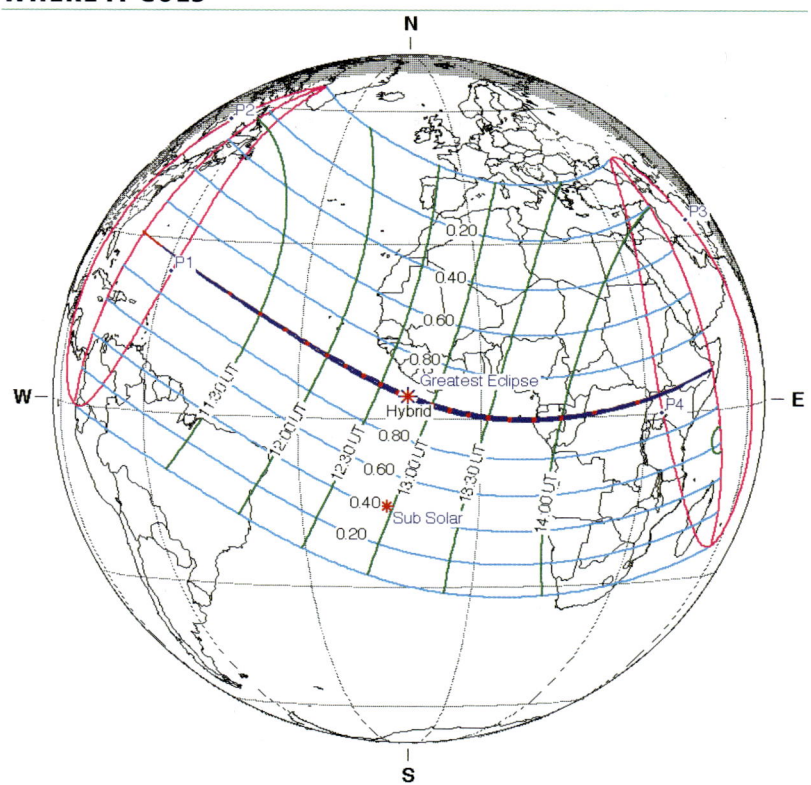

For an explanation of how a hybrid eclipse differs from a total eclipse see page 19.

THE ECLIPSE PATH The hybrid solar eclipse of 2013 starts as an annular eclipse in the Atlantic 1,000km east of South Carolina in the USA. The eclipse is annular for the first 400km of its track, after which it becomes total. Interestingly it passes through 0°, 0° (0° latitude, 0° longitude) before passing just south of the islands of Príncipe and São Tomé where evidence was gathered during the 1919 total eclipse to prove Einstein's theory of relativity. It enters Africa near Batanga on the Gabon coast, passing into the Congo, then the Democratic Republic of Congo (DRC), Uganda, Kenya, Ethiopia and finally Somalia at sunset. To witness a maximum duration of one minute and 40 seconds, you will need to be on a ship located at 3° 30'N, 11° 42'W; however, the weather prospects here are no better than on land. That being said, a ship can at least manoeuvre to find a hole in the clouds.

For a detailed map of the track see http://xjubier.free.fr/en/site_pages/solar_eclipses/HSE_2013_GoogleMapFull.html.

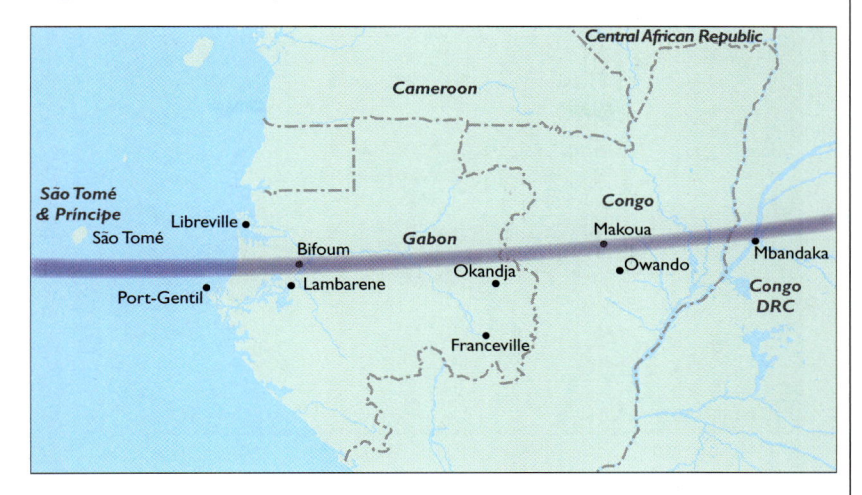

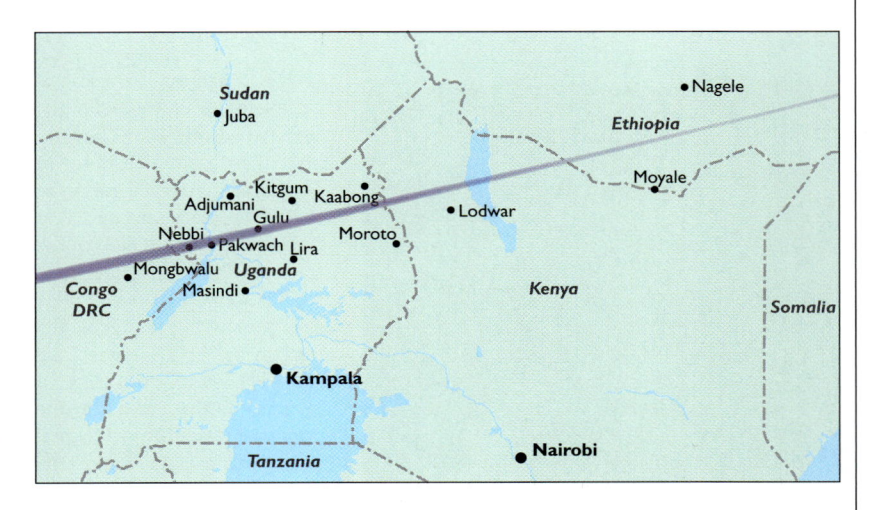

THE CLIMATOLOGY OF THE 2013 ECLIPSE TRACK

OVERVIEW The track of the 2013 hybrid brings an imposing meteorological challenge. Its beginning seems promising, in the relatively sunny skies beneath the subtopical highs of the Atlantic south of Bermuda, where it is unfortunately an annular eclipse. Enthusiastic eclipse travellers will therefore elect for a position further down the track, where the Moon covers the entire solar disc. Alas, as annular turns to total, the track moves out from under the favourable anti-cyclonic skies and into the influence of the Intertropical Convergence Zone (ITCZ). The ITCZ is a region where winds from the northern and southern hemispheres converge, squeezing the humid tropical air upwards, resulting in frequent heavy showers and thundershowers; and the ITCZ is particularly active over Africa in this season.

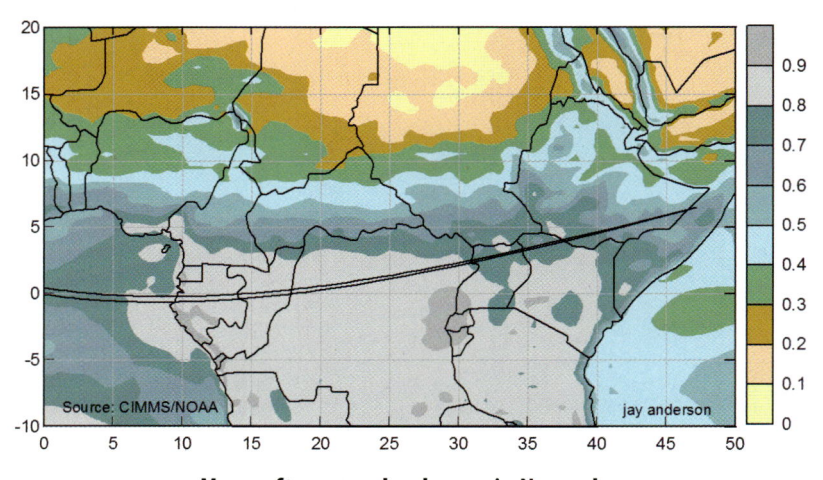

Mean afternoon cloud cover in November
Climatology report courtesy of Jay Anderson.

The Murchison Falls in northern Uganda lie near the 2013 hybrid eclipse track (AZ)

The eclipse track is completely over water for the first two-thirds of its length, and then comes ashore in equatorial Africa for the remainder of its journey across the Earth. Roughly 60% of the track is within the influence of the ITCZ, and therefore clear sky prospects are generally meagre. Three spots are relatively more promising: a position on the water near the point of maximum eclipse; a beachfront site where the shadow first comes ashore in Gabon; and at the end of the track, over Kenya and Ethiopia.

Satellite observations of global cloud cover measure an average cloudiness over the annular and early total portion of the track – at 60–70%. From about 15°N until the African coast, average cloudiness increases to 70–90% except for a zone of better weather south of

> **2013 HYBRID ECLIPSE FACTS**
> **Time of greatest eclipse** 12:46:28 UT
> **Max duration of totality** 1m 33s at latitude 3.537°, longitude 11.669°W
> **Altitude of Sun at greatest eclipse** 71°
> **Gamma** 0.327
> **Maximum width of path of totality** 58km
> **Saros Series** 143, member 23 of 72.

Liberia, where the eclipse reaches its maximum duration. In this region, satellite measurements of cloudiness drop slightly to 60–70%.

When the track leaves the water and crosses onto land, two additional cloud-making factors come into play. One is the greater heating of land surfaces compared to water; the other is a general increase in humidity as a result of the transpiration of water vapour by the jungle vegetation. Both contribute to the destabilisation of the atmosphere and the formation of convective clouds. The northern regions of the DRC have the greatest frequency of thunderstorms in the world.

Satellite measurements reveal a frequency of 80–90% cloudiness from Gabon across the DRC to the Ugandan border. Further east, over Uganda, Kenya and Ethiopia, the average cloud amount declines by 10–20%. Mountains in these three countries impose large variations on cloud cover, increasing or decreasing the background amount by about 10% locally depending on whether the observing site is on the windward or leeward slope.

While a certain amount of pessimism is inevitable given the climatology of central Africa in November, the weather station observations are less depressing. Ground-based measurements at Lamborini and Port Gentil in Gabon show that about one-third of the hours in November are sunny – possibly the best estimate of the probability of seeing the eclipse. Satellite observations are overly gloomy, as they record the amount of cloud rather than its transparency, relegating all clouds to the observational dustbin instead of just those that are opaque. Moreover, cloud cover under the ITCZ is extremely variable, as convective showers and thundershowers tend to develop in clusters that leave large areas of relatively sunny skies nearby.

In Gabon, eclipse seekers can take advantage of the proximity of the Atlantic Ocean and its slightly lower cloud cover. The country offers waterfront locations where the cloud climatology is influenced by the nearby cooler ocean waters and where a cloud 'advantage' of perhaps 5–10% may be realised. Such sites must be directly against the coast, as locations even a kilometre or two inland will have surrendered the advantage. The eclipse comes ashore in a relatively uninhabited part of Gabon and travel to the centre line will pose a challenge for larger expeditions.

Sunshine measurements from surface-based meteorological stations are in short supply across the DRC and Uganda, but Lodwar in Kenya reports a very encouraging 'percent of possible sunshine' of 74%. Lodwar is somewhat north of the eclipse track, but the on-the-ground measurement is confirmation of a pronounced decline in cloudiness towards the eastern end of the eclipse track. There is a penalty to be paid however, as the duration of the eclipse declines to less than 20 seconds across Kenya and to less than ten seconds in Ethiopia. For the experienced eclipse chaser, such a narrow and abrupt eclipse will come with a prominent presentation of the colourful innermost atmospheric layer of the Sun (the chromosphere) and a spectacular view along the axis of the Moon's shadow.

LOCAL ECLIPSE CIRCUMSTANCES

Location	Lat	Long	1st	2nd	3rd	4th	Alt	Dur
Max eclipse	3.537°N	11.669°W	11:04:24	12:45:40	12:47:19	14:29:38	71°	1m 33s
Gabon (coast road)	0.427°S	9.294°E	12:12:45	13:50:23	13:51:31	15:14:26	47°	1m 04s
Uganda	2.767°N	32.299°E	13:07:34	14:22:55	14:23:16	15:27:52	16°	0m 18s
Alia Bay	3.704°N	36.260°E	13:13:03	14:25:12	14:25:26	15:27:47*	12°	0m 11s

* After sunset. (Durations are corrected for the lunar limb profile.)

Predictions courtesy of Xavier Jubier.

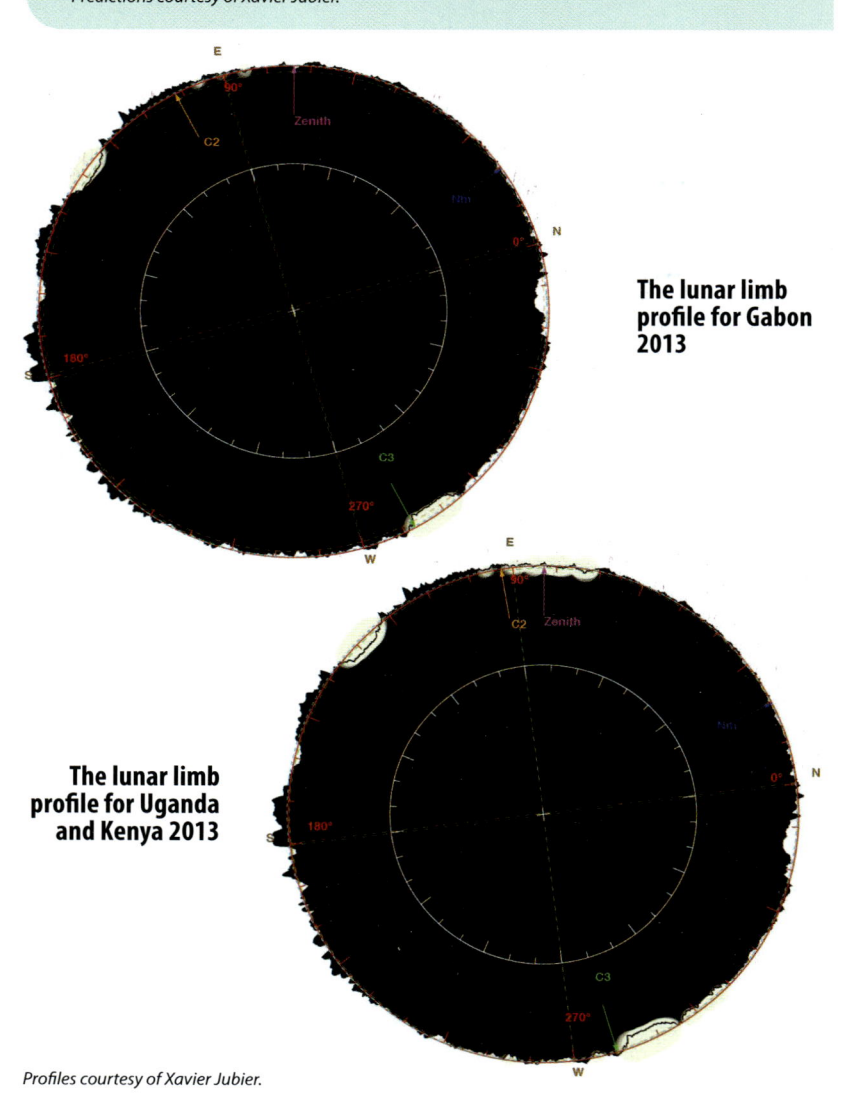

The lunar limb profile for Gabon 2013

The lunar limb profile for Uganda and Kenya 2013

Profiles courtesy of Xavier Jubier.

SKY CHART FOR 2013 During totality, planets and brighter stars should be visible. The chart below shows what to look out for.

Venus will be the most prominent object in the sky, visible many minutes before totality. This eclipse may also present a lot of people with their first ever glimpse of Mercury when it is very close to the Sun. Saturn is well placed too, also close to the Sun. Stars that should be visible include Arcturus, Antares, Spica, Vega, Altair and Deneb.

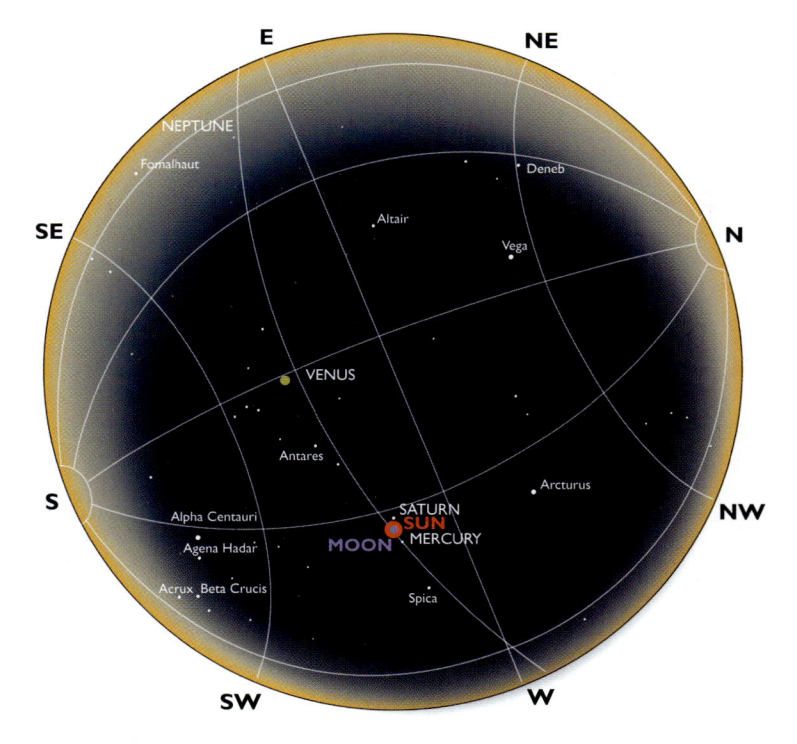

The chart above is for Gabon. For the Uganda/Kenya location the eclipse will be much closer to the horizon.

Sky chart courtesy of Xavier Jubier.

4

Annular Eclipses

INTRODUCTION

Annular solar eclipses are completely different from total solar eclipses. The Sun is never completely covered by the Moon so darkness never descends on the surroundings. Also, they can only be observed with the use of solar filters, and only with the naked eye if they are *very* close to the horizon, and even then, great care must be taken.

During a total eclipse there is almost no noticeable darkening until the Sun is 99% covered. During the annular eclipses of 2012 and 2013 in Australia, the Sun is never more than 86% covered. As such, hundreds of years ago when eclipses were not predicted, an annular eclipse could have happened without anyone being aware.

STAGES OF AN ANNULAR ECLIPSE

First contact: The moment when the edge of the Moon first encroaches on the Sun.

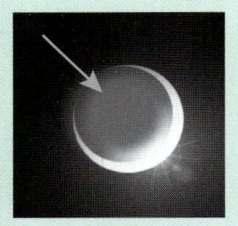

Second contact: The Moon is now completely inside the Sun.

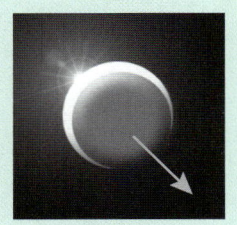

Mid-eclipse: The Moon doesn't cover the Sun completely.

Third contact: The Moon is just about to leave the Sun.

Fourth contact: The end of the eclipse.

An annular solar eclipse in progress at Dog Beach, San Diego; the image is a composite, for effect (SS)

Strangely, the longer the duration of an annular eclipse, the harder it is to see. In long duration annular eclipses the Moon is much smaller than the Sun and hence there is very little reduction in light. Another strange fact is that it is best to see an annular eclipse at either the northern or southern limit, rather than on the centre line. The reason for this is that you get extended duration Baily's Beads as the Moon skims the edge of the Sun. With a digital SLR camera it is even possible to photograph the eclipse without filters and see prominences, provided a very high shutter speed and low ISO are used. You should not look at the Sun through the viewfinder without a solar filter across the end of the lens. Frame and focus the eclipse with the filter in place, then, without looking through the viewfinder, remove the filter, and take the shot. Put the filter back on immediately to protect the lens and your eyes. The direction of the Moon's travel varies depending on your location. It will traverse from bottom right to top left in the USA in 2012, and from top to bottom in Australia in 2013.

THE ANNULAR SOLAR ECLIPSE OF 20/21 MAY 2012

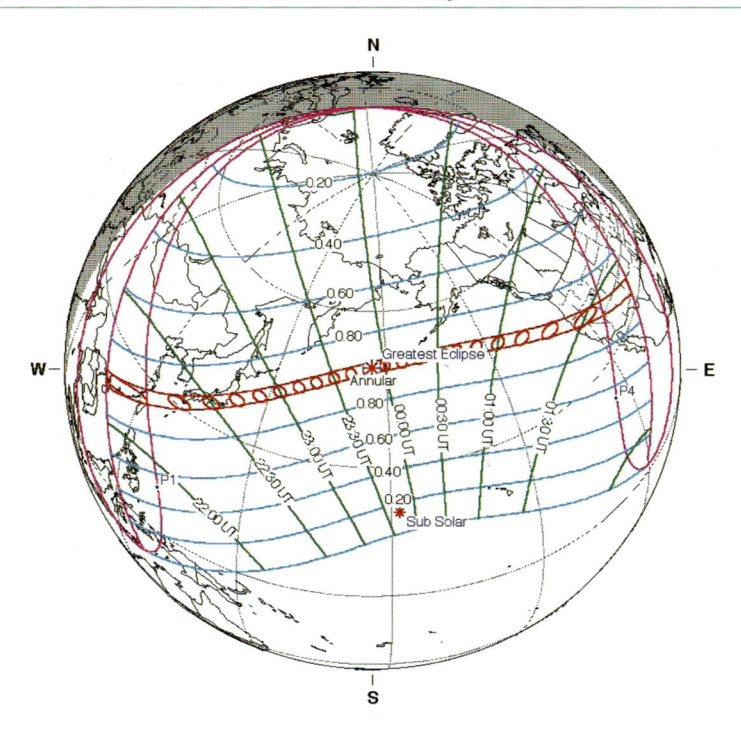

WHERE IT GOES

The eclipse path The path of the eclipse starts just to the west of Hong Kong, and then passes over Quanzhou in China. After leaving the mainland the eclipse just touches the northern part of Taiwan, crossing the Eastern China Sea before entering Japan. Here it traverses Tokyo and heads off across the Pacific to the United States of America. It passes north of San Francisco, and then into Nevada, Utah and New Mexico before coming to an end over the west of Texas.

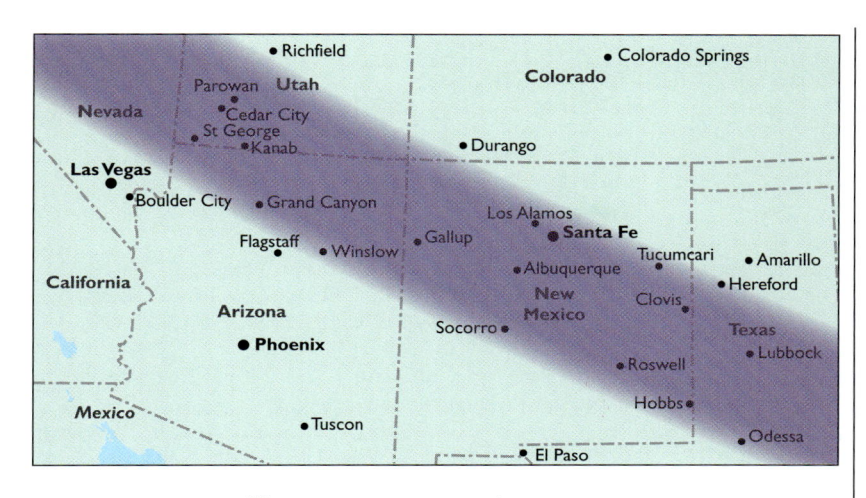

Albuquerque area – track near sunset

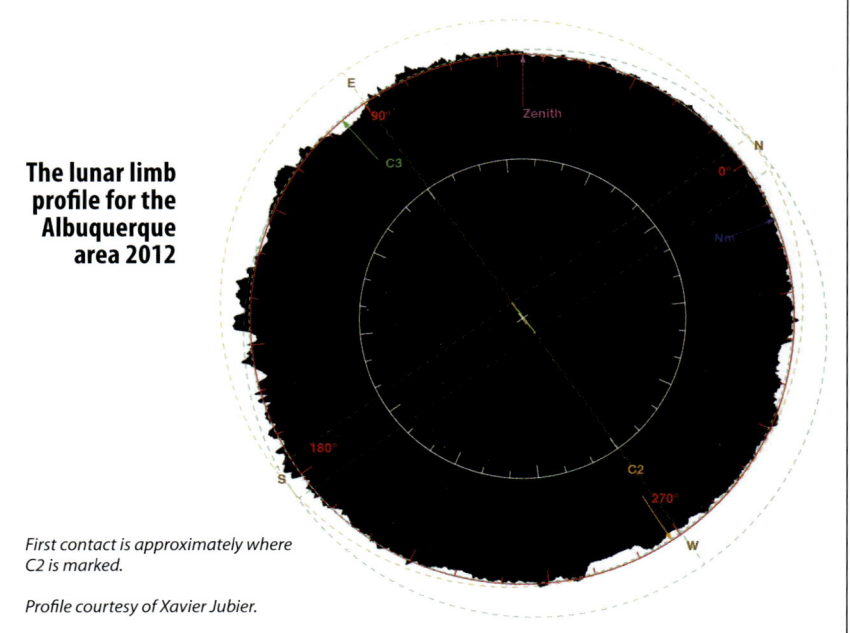

The lunar limb profile for the Albuquerque area 2012

First contact is approximately where C2 is marked.

Profile courtesy of Xavier Jubier.

2012 ANNULAR ECLIPSE FACTS

Time of greatest eclipse 23:52:47 UT
Max duration of annularity 5m 46s at latitude 49°05'N, longitude 176°17'E
Altitude of Sun at greatest eclipse 61°
Magnitude 0.9544; **gamma** –0.2695
Maximum width of path of totality 237km
Saros Series 128, member 58 of 73

LOCAL ECLIPSE CIRCUMSTANCES

Location	Lat (°N)	Long(°)	1st	2nd	3rd	4th	Alt	Dur
CHINA (20 May)								
Hong Kong	22.29	114.19E	21:08:10*	22:06:46	22:10:11	23:16:20	05°	3m 13s
Quanzhou	24.95	118.58E	21:08:27*	22:08:41	22:13:11	23:21:49	10°	4m 25s
Tokyo	35.44	139.58E	21:18:39	22:31:27	22:36:31	00:01:58	35°	4m 55s
USA (21 May)								
Albuquerque	35.11	106.66W	00:28:20	01:33:35	01:38:01	02:36:23*	05°	4m 26s
Sandia Peak	35.21	106.45W	00:28:15	01:33:27	01:37:49	02:36:10*	05°	4m 23s
Santa Fe	35.68	105.94W	00:27:46	01:33:10	01:36:45	02:35:24*	05°	3m 35s

San Francisco is not within the annular zone so has no 2nd and 3rd contacts. The time of maximum eclipse is therefore shown below:

Location	Lat (°N)	Long(°)	1st	Max	4th	Alt	Mag
San Francisco	37.73	122.43W	00:15:56	01:32:43 (mid)	02:39:58	19°	84⅔%

* Sun below horizon. Times above are all GMT.
Predictions courtesy of Xavier Jubier

THE ANNULAR SOLAR ECLIPSE OF 10 MAY 2013

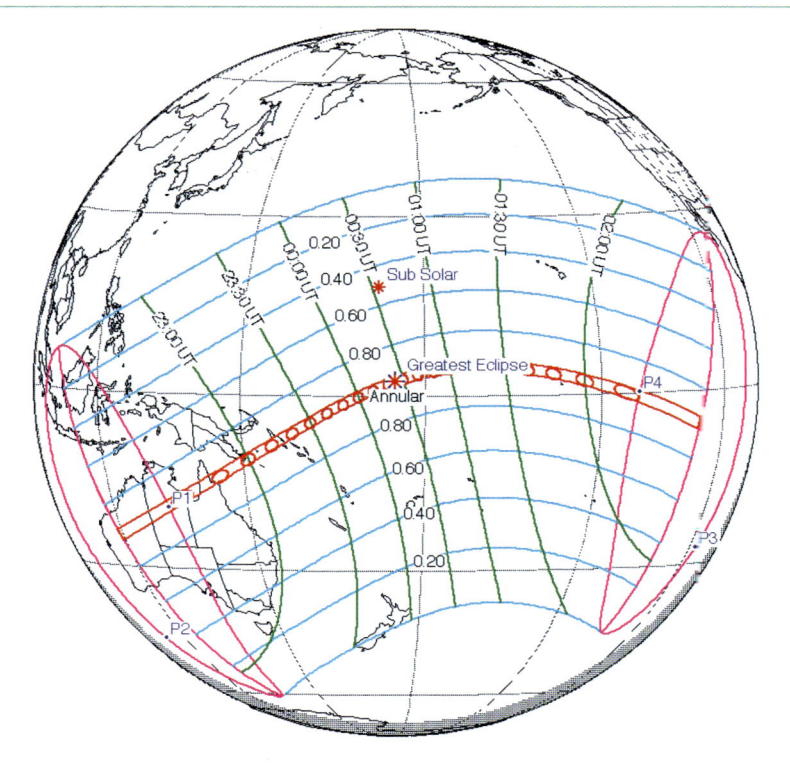

WHERE IT GOES

The eclipse path The path of the eclipse starts in the deserts of western Australia, crossing the 2012 total eclipse track in northern Queensland. It does not traverse any major towns in Australia before heading across the sea to touch the extreme eastern tip of Papua New Guinea and parts of the Solomon Islands. Once over the Pacific it crosses Ocean Island, Bairiki, Fanning Island, and just misses Kiribati (Christmas Island).

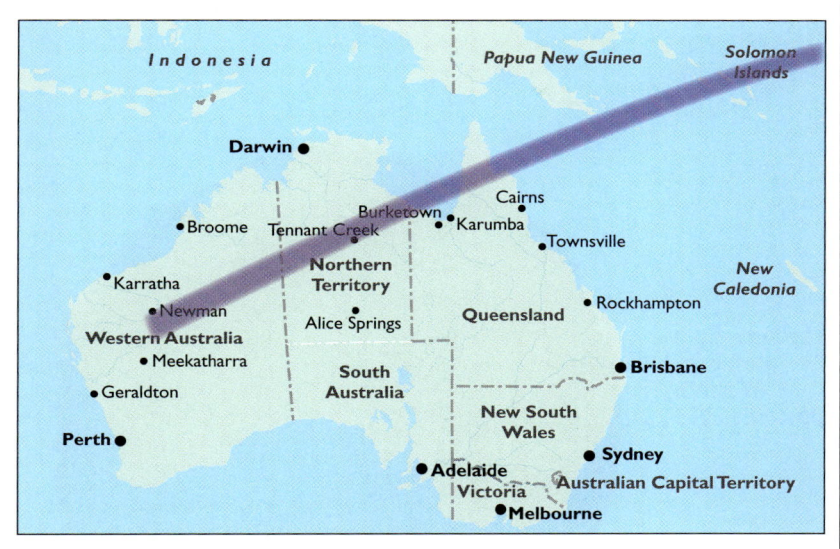

LOCAL ECLIPSE CIRCUMSTANCES

Location	Lat(°S)	Long(°E)	1st	2nd	3rd	4th	Alt	Dur
AUSTRALIA								
Newman	23.36	119.73	21:28:15*	22:31:32	22:33:23	23:45:54	01°	1m 51s
Tanami	19.97	129.72	21:25:26*	22:33:21	22:36:16	23:56:10	11°	2m 40s
Desert Queen	22.46	122.26	21:27:16*	22:32:02	22:33:16	23:47:53	03°	1m 14s
Tennant Creek	19.65	134.19	21:25:10	22:35:45	22:38:50	00:02:44	16°	3m 05s
PAPUA NEW GUINEA								
Alotau	10.33	150.46	21:32:23	22:57:20	23:01:04	00:46:38	38°	3m 25s
SOLOMON ISLANDS								
Gizo	08.10	156.84	21:39:04	23:12:07	23:15:01	01:10:33	48°	2m 39s

The following Australian locations not within the annular zone so have no 2nd and 3rd contacts. The time of maximum eclipse is therefore shown below:

Location	Lat(°S)	Long(°E)	1st	2nd	3rd	4th	Alt	Mag/
Cairns	16.89	145.78	21:27:58	22:48:43 (max)		00:27:04	29°	88%
Karumba	17.49	140.84	21:26:04	22:43:00†		00:15:45	23°	88%

* Sun below horizon. (Durations are corrected for the lunar limb profile.)
Predictions courtesy of Xavier Jubier.

2013 ANNULAR ECLIPSE FACTS

Time of greatest eclipse 00:25:13 UT
Max duration of annularity 6m 03s at latitude 02°13'N, longitude 175°28'E
Altitude of Sun at greatest eclipse 74°
Magnitude 0.9544; **gamma** −0.2695
Maximum width of path of totality 176km
Saros Series 138, member 31 of 70

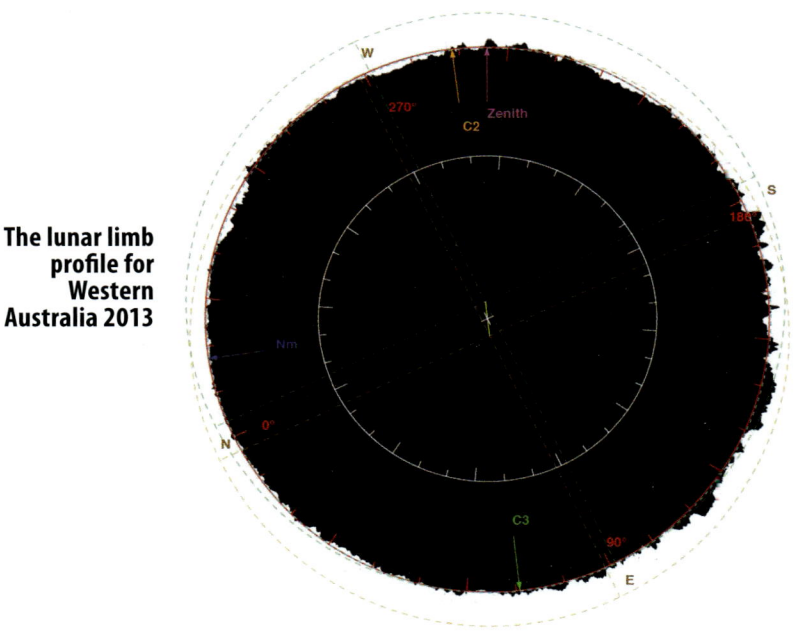

The lunar limb profile for Western Australia 2013

First contact is approximately where C2 is marked.

Profile courtesy of Xavier Jubier.

5

The Venus Transit of 6 June 2012

INTRODUCTION

Transits of Venus are very rare events and usually occur in pairs, with one four or eight years apart from the other. The previous occurrence was in 2004 when the whole transit was visible from Britain. The next transit of Venus will not occur for another 105 years until 2117, so you can safely say that if you miss this one you will *never* see another. (See *Further Information*, page 48.)

WHERE IT CAN BE SEEN

It takes over six hours for Venus to cross the Sun, and the map diagram on page 34 shows that the whole transit is only visible from eastern Australia, New Zealand, eastern Asia, Japan and the northwest Pacific. It is also visible in its entirety in the north polar regions, including Iceland, where the Sun never sets in summer. Most of South America and western Africa, Portugal and western Spain will see nothing. The remaining areas, including Britain, will see either the beginning or the end of the transit.

LOCAL TRANSIT CIRCUMSTANCES

Location	Sunrise	Sunset	I	II	M	III	IV
London	03:45	20:13	-	-	-	04:37	04:55
Cardiff	03:58	20:26	-	-	-	04:37	04:55
York	03:35	20:32	-	-	-	04:37	04:55
Exeter	04:03	20:23	-	-	-	04:37	04:55
Edinburgh	03:30	20:54	-	-	-	04:37	04:55
Inverness	03:23	21:09	-	-	-	04:37	04:55
Lerwick	02:46	21:22	-	-	-	04:37	04:54
Belfast	03:51	20:55	-	-	-	04:37	04:55
Reykjavík*	03:11	23:44	22:04	22:21	-	04:36	04:54
Los Angeles	12:40	03:02	22:06	22:24	01:25	-	-
Hawaii	14:19	05:04	22:10	22:28	01:26	04:26	04:45
Tokyo	19:25	09:54	22:11	22:28	01:30	04:30	04:47
Sydney	19:26	06:53	22:16	22:34	01:30	04:26	04:44
Wellington	18:02	04:58	22:16	22:34	01:29	04:26	04:44

*Reykjavík: the Sun sets for over 3 hours from 23:44 to 03:11 missing mid-transit.
M = time of mid transit.
Times are UT

33

Courtesy of Fred Espenak, NASA

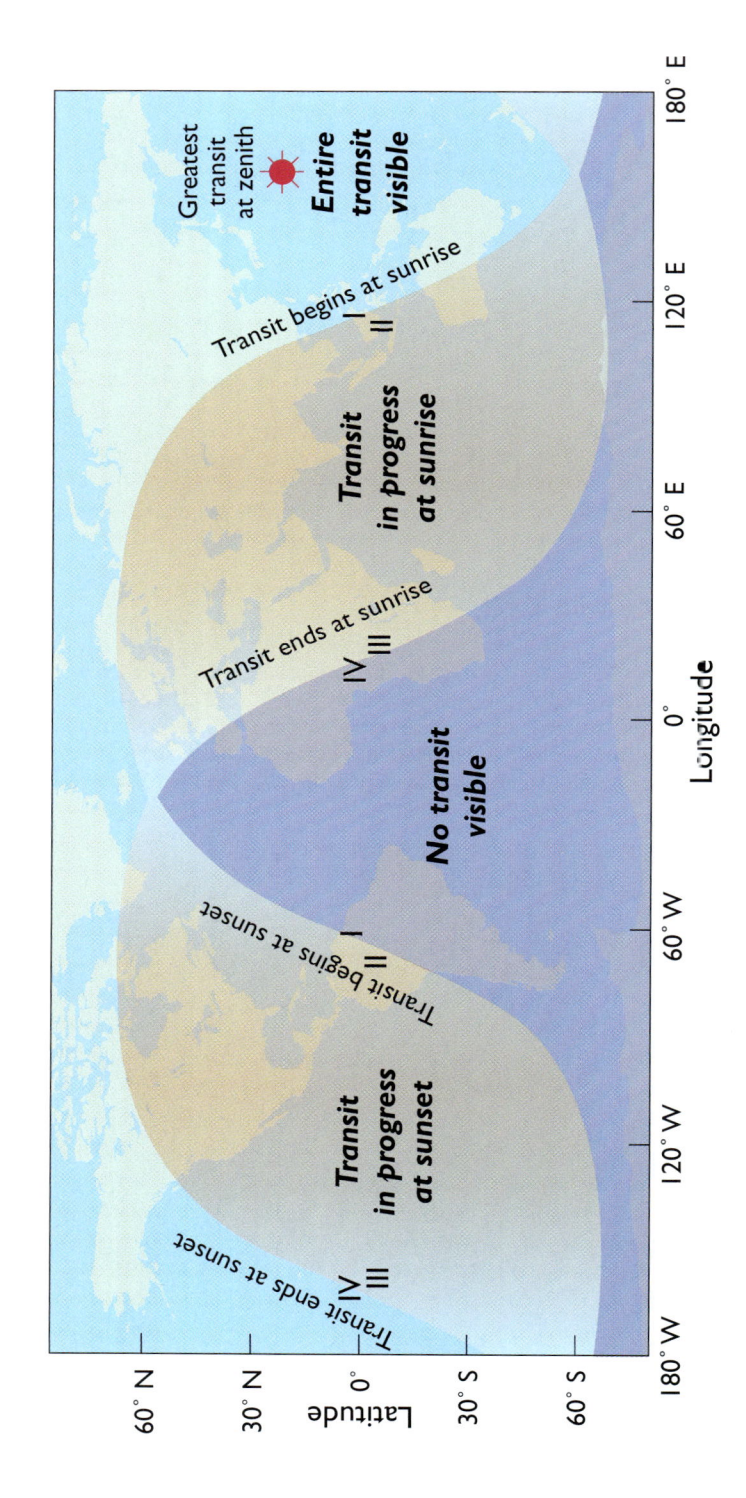

Venus transits must be viewed by projection or using solar filters which can be obtained from many places (see *Further Information*, page 48). For information on viewing the transit, refer to *Viewing eclipses safely*, page 36. Note that Venus is large enough for you to see it by simple projection using a solarscope.

The diagram below shows where on the Sun's disc Venus will appear. Note that if you are using a telescope the image can be reversed or inverted.

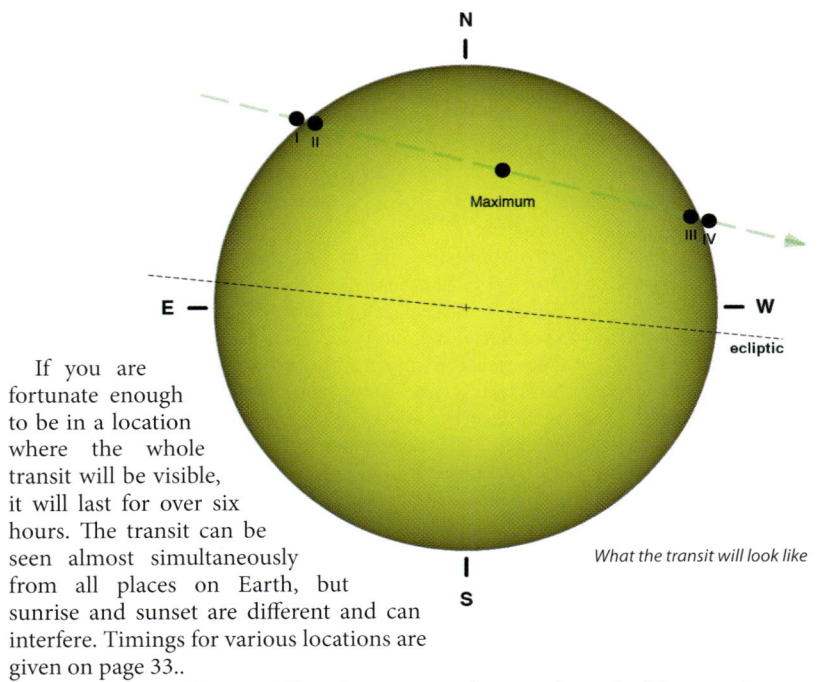

What the transit will look like

If you are fortunate enough to be in a location where the whole transit will be visible, it will last for over six hours. The transit can be seen almost simultaneously from all places on Earth, but sunrise and sunset are different and can interfere. Timings for various locations are given on page 33..

As you can see, Britain will see just over one hour at the end of the transit (times are GMT so add one hour to get BST). Interestingly, at Reykjavik the Sun sets for over three hours from 23:44 to 03:11 missing mid-transit, but it does see its beginning and the end.

6

Planning, Preparation and Photography

HEALTH AND SAFETY

GENERAL HEALTH Each of the eclipses traverses completely different climatic and geographical regions, so you should consult country guides and specialist health guides for advice on immunisation.

Some first-time visitors to countries not on the normal tourist trail initially experience a degree of culture shock, unless they are cocooned inside first-class hotels and whisked through the countryside in luxury coaches. You might find it hard to cope with the noise, dirt and poverty of the cities, and with the simple standards of living in rural areas. More importantly, think twice before the eclipses lure you off the beaten track unless you know from experience how to prepare for such ventures. In every one of the eclipse countries there are regions where simple mistakes like forgetting the drinking water, getting stuck in sand or taking the wrong clothes, can turn into disasters. The usual means of recovery – shops, telephones, medical facilities, other vehicles – may simply not be there.

VIEWING ECLIPSES SAFELY It is dangerous to look at the Sun at all times except for the brief period of totality, when it is fully covered by the Moon. Some eye experts argue that you should never look at an eclipse, even during totality, because it is all too easy to make a mistake, such as looking too soon, and cause permanent damage to your eyes. Astronomers and other experts say sensible adults should easily be able to follow the basic viewing rules to witness one of the great sights of nature.

If you stare at the Sun you are looking directly at a huge, thermonuclear explosion spewing out ultraviolet and infrared rays. Evolution has honed us to find it painful to look at the ultraviolet rays – if your gaze accidentally settles on the Sun, discomfort quickly forces you to look away. But it is the infrared, rather than the ultraviolet, which does most of the damage. Infrared is heat, and heat cooks the delicate tracery of blood vessels at the back of the eye. If your retina is damaged you will become blind at the spot where images are focused, making it hard to read or recognise people's faces. Nothing in medical science can reverse this.

During a transit or when the Sun is partially eclipsed or annular there may be insufficient ultraviolet rays to make it painful to gaze at the Sun – but there will definitely be enough infrared to damage the eye. Even at 99% eclipse the rays from the Sun are 4,000 times stronger than those from the Moon. That is why watching eclipses can be dangerous.

The one time when it is safe to watch the eclipse is during totality – the few minutes when the Sun is totally covered by the Moon. The Sun is then only as bright as the full Moon and cannot cause damage. For the rest of the eclipse you must not look with the naked eye. It takes mighty screening power to block the Sun's rays and the only suitable materials are specially made, and generally coated with a thin layer of metal such as silver or aluminium. Often the material is aluminised mylar or Baader AstroSolar Safety film; welder's goggles with a rating of 14 or higher are also suitable. Nothing else is safe, and that includes sunglasses, smoked glass, compact discs, exposed black-and-white film, and the sun caps or solar eyepieces provided with binoculars and amateur telescopes (you need special metal-coated filters for these).

For covering your camera lens, telescope or binoculars you will need filters you can cut to shape. A4 sheets are available from:

Broadhurst Clarkson & Fuller
www.telescopehouse.co.uk

SCS Astro www.scsastro.co.uk

Alternatively, simply type Baader AstroSolar into your favourite internet search engine. Baader filter sheets come with instructions detailing how to make a cardboard mount to hold the filter material, and jolly effective it is too.

You can order eclipse glasses from many companies, but most tour companies will provide them for their clients. It is still advisable to purchase your own to share with local people. Some well-known suppliers are:

3-D Images Ltd 020 8364 0022; www.3dimages.co.uk/eclipse_glasses.html
Assistpoint Ltd 4 Allendale Rd, Barnsley, South Yorkshire S75 1BJ; www.assistpoint.co.uk. Available for purchase via their website or for 5 viewers send a cheque for £10.00 (10 viewers £19; 25 viewers £45) made payable to Assistpoint Limited.

Rainbow Symphony 6860 Canby Av, #120, Reseda, CA 91335; 001 818 708 8400; www.rainbowsymphony.com

Test your glasses beforehand against a bright light source such as a reading lamp. If there are any pinhole pricks throw the glasses away. Ideally take several pairs in case one becomes damaged – keep them where they cannot get scratched.

Watch the stages of the eclipse through the approved glasses but, even with this protection, do not stare for longer than half a minute. This is the belt and braces approach – a tiny hole you have missed is not likely to cause harm (damage from gazing directly at the Sun has been reported with less than a minute's viewing). When the diamond ring appears you can remove the glasses, replacing them when it returns at the end of totality.

Eye experts worry that the slightest accident when following the above protocol could lead to permanent damage. They fear in particular, glasses slipping from a child's face or an adult forgetting the rules amidst the tremendous excitement of the eclipse. They point out that the only totally safe way to view an eclipse is through a pinhole projector. This can be made by piercing a hole in a piece of card and viewing, on another piece of card, the light that shines through it. Do NOT look through the pinhole!

A new way of observing the partial phases is by use of a Solarscope. This device comes as a flat-packed box with a metal enclosed lens. It allows several people at once to look at the Sun's projected image. Once constructed it becomes a stiff cardboard structure darkened on the inside. It has a lens at the end of the metal fitment plus a small adjustable convex mirror that reflects the Sun's image back into the box. Solarscopes cost approximately £35 and are available from many sources such as:

Broadhurst Clarkson and Fuller
www.telescopehouse.co.uk
Green Witch www.greenwich-observatory.co.uk

Widescreen Centre www.widescreen-centre.co.uk/teledirect/

Viewing a total eclipse in northern South Africa; it is vital to use a viewer when looking at the partial eclipse (SS)

LOCAL EYE SAFETY Ophthalmologists say there is bound to be an epidemic of eye damage among people who are not aware of the dangers of viewing the eclipse. For once, a majestic work of physics conducted in the biggest laboratory of all will happen right in front of many isolated people. The problem is that if these people watch the drama unprotected, they could damage their eyesight for life. Tourists can act for good or evil here. Your presence may encourage people to stay outside and watch the eclipse instead of running indoors which is the instinctive response. Alternatively, you can be a source of information (and viewing glasses) so that those who would otherwise have stared long and hard at the eclipse learn how to view it safely.

Several groups are working to help local people make the most of the eclipses while safeguarding their sight. Below are some ways you could help:

- Take several viewers with you.
- Even if you have only one viewer with you, lend it to those without them. You will find you want to look at the partial eclipse only briefly, once every ten minutes or so. One viewer can therefore serve several people.
- Find out the official eye safety advice in the country you are visiting (if any) and try to work with this rather than confusing people by contradicting it.
- Remember the golden rule: do not look at any stage of the partial eclipse except through specially made viewers, projection or a solar scope.
- The primary method of viewing the eclipse is likely to be smoked glass. This is very dangerous (though better than watching with the naked eye). Offer the use of your viewers instead.

WHAT TO TAKE

Eclipse viewing glasses are vital (see *Viewing eclipses safely*, page 36). If you plan to take photographs you need to give some time and thought to equipment (see *Photography*, opposite). You may also want to take binoculars with the appropriate solar filters. A tape or digital recorder might seem like a strange thing to bring to a spectacle but the responses from the people around you will be tremendously vocal, and you can add your own adjectives and descriptions as you watch. (Alternatively, some people pre-record minute-by-minute instructions about what they should look for and when, and then play the recording back when the eclipse starts.) Thin sheets of cardboard from which to make a pinhole camera may be useful; they can also be used to make holes to project tiny crescent Suns onto the ground. In addition, a tripod, a seat or rug, a torch, protection from the Sun, mosquito repellent (for when the insects emerge in the twilight) and a water bottle can come in handy.

PHOTOGRAPHY

> I well remember that I wished I had not encumbered myself with apparatus, and I mentally registered a vow, that, if a future opportunity ever presented itself for my observing a total eclipse, I would … devote myself to that full enjoyment of the spectacle which can only be obtained by a mere gazer.
> Warren De La Rue, on the total solar eclipse in Spain of 18 July 1860

If this is your first total solar eclipse, consider first whether you want to be distracted during the precious few minutes of totality by the need to focus cameras and adjust settings. You may absorb the moment better if you forget about photography and ask fellow travellers for copies (in my experience they will be flattered and only too pleased to provide these), or buy professional photographs later – there will be many of them and they will be superb.

DIGITAL OR NON-DIGITAL CAMERAS? Nowadays digital SLRs are available for less than £400, and you can buy used models from eBay for far less. Perfectly acceptable compact digital cameras cost less than £200. This doesn't mean that non-digitals are of no use; on the contrary, they continue to give excellent results although digital cameras are very tolerant of extreme exposures, and give instant feedback on whether you are doing it right (or wrong). Most of the advice given here applies to both types, and if the advice differs between the two it will be stated explicitly.

COMPACT CAMERAS Compact cameras are ideal for capturing the surroundings during totality, but because of the focal length of their lenses, the eclipse itself will be but a tiny dot in the distance (even on full 3–5x zoom). Instead I recommend that you turn off the flash (or better still cover it with opaque tape). This is very important for two reasons – first you will annoy other eclipse watchers, secondly you want to record the reducing light levels and the flash will negate this. Proceed as follows: mount the camera on a tripod (because the long exposures will induce camera shake and give blurred images); you won't need a filter as you are not pointing the camera at the Sun. Point the camera horizontally (not up at the Sun) and put in a new film or memory card that will allow at least 30 exposures. See if the camera will allow you to set the exposure and shutter speed manually. If so, set it to 1/50s and with the widest aperture possible. Try not to look through the viewfinder, use the screen instead. Now take shots as follows:

Time to go until second contact	Frequency of photo every
5 mins	1 min
1 min	20 seconds
Totality	10 seconds

Don't try to fiddle with exposure settings during totality, but do turn the camera around on its tripod to change the view. If you can overlap the images during totality by 20% and have a digital camera, there are many panorama-stitching programs available when you get home to create a truly memorable scene.

SLR CAMERAS For totality itself you need to use a single lens reflex (SLR) camera or a camcorder, and the text below gives advice on using these.

Filter Don't forget that gazing through your camera viewfinder at the Sun is dangerous – the rays will fry your retina as well as the electronics inside your camera. You must have a neutral density filter that cuts out light and heat by a factor of 100,000. This translates to an ND5.0 filter. You must use it even when the Sun is 99% covered, removing it just before Baily's Beads appear.

Earlier it was mentioned how the Baader AstroSolar filter A4 sheets come with instructions on how to make your own camera and binocular filters (see page 37 for further information). Large filters for telescopic work can be ordered from Thousand Oaks Optical (*www.thousandoaksoptical.com*). Be sure to give lots of lead time for your order.

Lenses Before choosing the lens to attach to your SLR it is important to realise that current budget digital SLRs produce an image 1.6 times larger with the same

Broken-ring eclipse composite taken while flying at 40,000ft in a Cessna Citation aircraft (GS)

focal length lens when used on a 35mm film camera. There are full-frame digital cameras available from Canon, Nikon, Sony, etc, but these are very expensive, where lenses will be equivalent. So a 200mm lens attached to a digital camera will produce an image equivalent to 1.6 x 200 = 320mm lens attached to a non-digital SLR (making the Sun appear 1.8mm actual size). Bracketed values below are the rough equivalent lens sizes for digital cameras.

In order to get a 2mm image of the eclipsed Sun (ie: just to discern it as a disc) you will need at least a 250mm lens (160mm). However, for eclipse close-ups you will need a lens (or telescope) with a focal length of 500mm (300mm) or more. A 500mm (300mm) lens will yield a 4.5mm diameter image on the film or chip. Better still would be a 1,000mm (600mm) lens to produce the largest image possible while leaving space to accommodate the inner corona, chromosphere and prominences around the Sun. Do not use longer focal lengths unless you want to image the Sun's disc without the corona. To work out the size of the image on the resulting photograph, divide the focal length of the lens (in mm) by 110, or 70 for a digital SLR. Note that you can also buy lens multipliers, and a 2x multiplier will make your 250mm lens into a 500mm lens (although you will lose light collecting power).

You can also buy small portable refracting telescopes that are ideally suited to eclipse work. I use a William Optics 66mm (350mm focal length) telescope, but there are many other suitable models. If you wish to follow this route you are advised to approach a reputable telescope specialist as it is critical for the optics to be of good quality. Please think twice before buying a telescope from anyone else.

Tripod, cable release and sequence control The final essential piece of equipment is a tripod. Lightweight tripods frequently cause blurred eclipse photographs but are better than no tripod at all. The tripod legs should not be extended more than halfway. Try adjusting the height so that you can easily reach the camera controls while sitting on a chair, thereby minimising the vibrations you transmit. You can further decrease vibrations by suspending a weight under the tripod – rocks or sand in a sack. Also use a cable release to stop camera shake when you press the shutter. Some digital cameras have cable releases that are programmable, and many digital SLRs can be controlled by a laptop computer so you can pre-determine every shot and sequence (however, without a driven camera mount, the Sun will drift across the camera's field of view quite quickly, and you will need to keep adjusting the position of the camera). You may also find a right-angled eyepiece adaptor useful (but rather expensive) as most eclipse locations have the Sun high in the sky. If you want to automate the whole set-up you will need a driven camera mount, and the AstroTrac is ideally suited to this. Combined with a programmable cable release you need not take a laptop with you, but you will need a 12v power supply (AstroTrac supply an 8xAA battery holder for this).

How to image using an SLR camera There are two key factors that will determine the success of a close-up eclipse image: focus and camera shake.

Focus is most difficult to cope with and requires practice. If you are using a standard lens, it is not simply a matter of setting the lens to ∞ (infinity). Practise at home before you go, and spend some time during the partial phases fine-tuning the focus, as this can change as the camera and lens heat up. Many modern digital SLRs now have 'live view' whereby you can look at the image on the screen at up to 10x magnification.

Details of how to minimise camera shake are detailed above, but there is also the option of using 'mirror lock up' ('live view' will do this for you) and ensuring the camera is set to the fastest possible shutter speed.

Total solar eclipses are among the most tolerant things to photograph. Provided you have a large enough lens and a tripod, every shot you take will show something of interest.

Simple approach Put the SLR on a tripod and preferably attach a cable release, set the camera to automatic exposure and make sure the flash is off. The choice of film (or the ISO setting on a digital camera) is not crucial, but I have found when using a tripod that ISO 800-1600 will allow you to use the fastest shutter speed to minimise camera shake. You must also turn off the auto-focus as the low light conditions usually prevent this from working. Spend some time during the partial phases finding the precise focus. You will probably over-expose the inner corona and prominences but you will produce reasonable pictures. This approach has the great advantage of being simple and leaving you time to watch the eclipse properly.

For those first-timers who are serious about wanting to photograph the event, please bear in mind that a total eclipse is intensely emotional, and an awe-

ECLIPSE CHASERS

You've heard of trainspotters – well, there are people whose sole aim in life is to go to the ends of the Earth (literally) to see a total solar eclipse. These people have names such as eclipse chasers, ecliptomaniacs, eclipsoholics and umbraphiles. We (I admit that I class myself as one) take total solar eclipses very seriously. For example, I use total solar eclipses to plan my holidays. In 1988 I went to Sumatra, 1990 to Finland, 1991 to the Baja Peninsula in Mexico, 1994 to Peru, 1997 to Siberia, 1998 to Antigua, 1999 to Cornwall in England, 2001 to Madagascar, 2006 to Libya, 2008 to the Gobi Desert, 2009 to Wuhan in China, and 2010 to Tahiti. In just over three years (2015), I will be able to see the next total eclipse to traverse British waters – this eclipse passes right next to the tiny island of Rockall on 20 March. What I am really waiting for is over six minutes of totality in Luxor, Egypt, in 2027, and of course the next total solar eclipse to cross the USA in 2017.

Umbraphiles will go to any lengths to find the best location to maximise the chance of seeing totality, and for the March 1997 eclipse most tours went to Mongolia. I decided that the eclipse would be too short and low on the horizon when seen from Mongolia, but the problem was that maximum duration and altitude was at latitude 60°N in northern Russia – inaccessible in March and unbelievably cold. I chose Chita in Siberia near Lake Baikal as the best bet. To cut a long story short, when I arrived in Chita I was walking past the station and who should get off the Trans-Siberian Express but Glenn Schneider, adding to other unplanned encounters with him in 1988 and 1991. According to my survey, Glenn currently leads my website league table highlighting the world's umbraphiles, having seen 29 total solar eclipses, and has been in the Moon's umbral shadow for over 79 minutes. I rate a mere 28th in this table (see my website: *www.clock-tower.com/eclipse.htm*).

Glenn is renowned for his exploits, which include a hike up a mountain on Atka Island in the Aleutians to see the 1990 eclipse. However, his most extraordinary endeavour was on 3 October 1986. A solar eclipse was due to happen but unfortunately it was only to be annular from the surface of the Earth – and an annular eclipse is no good for an umbraphile as the Moon is not big enough to cover the Sun completely. It just so happened that the Moon's umbral shadow (the only place in which a total solar eclipse can be seen) stopped 12km above the

inspiring spectacle. You will almost certainly be overcome by the sight and will probably not be able to achieve what you set out to do. Nevertheless the golden rule is preparation. First, decide what kind of pictures you want – evocative scenes of the Sun fringed by trees and people, or a close-up of the flaring prominences? Try out all new equipment and rehearse the procedure at home (the Sun will conveniently be there to practise on). This way everything has a higher chance of working perfectly at the crucial moment. To simulate totality, wait until half an hour after the Sun has set and practise changing the shutter speed from shortest to longest without a torch.

If you are photographing the partial eclipse you should calculate the shutter speed and aperture appropriate to your solar filter and telephoto lens by setting up the apparatus on a sunny day at home. Also, if your camera is digital and has a RAW setting, you should use this. Do make sure that you are using the highest resolution setting your camera has. You should also consider using the 'mirror lock up' feature if your camera allows this, to minimise camera shake. Again, practise

Earth's surface. A group of nine umbraphiles led by Glenn arranged to charter a Cessna Citation aircraft to fly up into the arctic air near Greenland and rendezvous with the Moon's tiny shadow, which was probably only a few hundred metres across. If their calculations were correct, and the pilot could fly accurately, they would witness a fleeting glimpse of totality, maybe only one or two seconds, but long enough to notch up another total eclipse.

This feat was successful and, as soon as totality ended, Glenn and his team looked down at the cloud tops and saw a cigar-shaped shadow of the Moon racing off into the distance, secure in the knowledge that they were the only living creatures to see this totality. To read more about this amazing feat, visit http://tinyurl.com/cjsu4.

There can be problems if you are a fanatical umbraphile. I recall that in July 1991 we were relaxing in luxury at a beach hotel in San Cabo at the southern tip of the Mexican Baja Peninsula, waiting for an eclipse that was to last six minutes and 43 seconds. It was to be the longest remaining eclipse in the lifetime of everyone on this planet. A group of umbraphiles amongst us had previously calculated that a 50km trip into the desert would yield an extra 11 seconds of totality, so they set off at dawn armed with kit to allow them to survive many hours with no shelter, with the midday Sun directly overhead.

Meanwhile we sat on the beach, occasionally taking half a dozen steps into our hotel rooms to collect another beer from the fridge. We saw a magnificent, full-duration eclipse in total luxury – unaware that the group that had set out earlier had a cloud drift in front of the Sun for over a minute.

The fanatical umbraphile will be armed with a laptop computer and global positioning system, as well as a detailed chart of the Moon's valleys and mountains. Using these, it is often found that maximum totality can be some way away from the centre line, allowing a valuable second or two to be added to the viewing log.

The best locations are frequented by thousands of umbraphiles, which in turn attracts the media, local dignitaries and others. This prevents umbraphiles from making exaggerated claims about their times under totality.

If you succeed in seeing totality I assure you that you will want to see it again – and then I'm afraid *you* will have become an umbraphile.

all this at home on the Sun many weeks before the eclipse. If you are using a digital camera, take the opportunity to buy a new card for the camera, and use this only for the eclipse.

More complex approach For this, set the camera to manual and make sure that RAW (or RAW+JPEG) mode is set. While still at home practise by pointing the camera at the Sun (using your solar filter) and set the aperture to f/8. Take a frame for every shutter speed from 1/4000s (or whatever is your shortest) to 1/2s, record the settings and view the results to see which works best (if you don't have a digital camera it's still worth using up a whole film to check, but do ask the processing company to develop all the exposures regardless). Bear in mind that if the sky is hazy at eclipse time you will need longer exposures to compensate. This approach allows you to practise getting the precise focus too.

Prior to totality Before the eclipse begins, set the camera to RAW (or RAW+JPEG), fit the filter, set the focus to infinity (some lenses focus beyond infinity and you may have to experiment to find precise focus), switch off auto-focus, switch to manual, turn off the flash and set the camera to f/8 and the shutter speed you determined by experimenting at home (f/8 is a good compromise allowing the use of fastest shutter speeds, minimising camera shake and yielding the best results across the whole frame). Do not hunt through the sky for the Sun without the filter fitted. Bear in mind that the partial phases last roughly an hour, so the Sun will move across the field of view and the camera will need occasional positional adjustment. Don't forget the cable release, if you have one.

Depending on how many exposures you have available, take one exposure at first contact, then one every five minutes (probably 12 exposures). Keep noting which way the Sun travels in the viewfinder so that when totality arrives you have the Sun positioned so that it remains in the viewfinder for up to two minutes (you do not want to have to adjust the position during totality). If you are using a focal length of more than 400mm and totality lasts more than two minutes, you will almost certainly have to move the camera at least once during totality itself.

It is recommended that one minute prior to the second contact (and the resulting diamond ring), you locate the Sun in the viewfinder (with the filter still on) in the position determined in the last paragraph (so in two minutes it will not drift out of the other side), then (and only then) carefully remove the filter so as not to move the camera.

During totality The difficult bit is determining the shutter speed for the build-up to totality. You don't want to have to change settings more often than you need. It will be dark and you may not even be able to see the dial or screen! So assuming you have set the ISO to 800 and the aperture to f/8, I suggest that you set the shutter speed to 1/2000s for the diamond ring and Baily's Beads. Using the cable release, fire off shots every two seconds (no more frequently in order to give the camera shake time to damp down). When all beads are gone and totality is upon you it becomes so dark that you will need to immediately increase the exposure successively in steps from 1/1000s to 4s depending on what phenomenon you want to capture. Each eclipse phenomenon has a different brightness value so exposures will vary according to what you want to photograph. For example, to capture the solar prominences and chromosphere with the suggested f ratio and ISO you need an exposure time of about 1/1000s. But to capture the extreme outer corona the exposure time should be as long as 4s (yes, four seconds); this should also show details on the Moon illuminated by earthshine! The solution is to bracket your exposures. Using the technique to change just the shutter speed you have practised at home in the dark, proceed as follows once the diamond ring has gone and totality is in progress:

- Start with a shutter speed of 1/1000s exposure, wait 5 seconds
- Change the shutter to 1/500s exposure, wait 5 seconds
- Change the shutter to 1/250s exposure, wait 5 seconds
- Keep doing this until the shutter speed is 4s

This will have occupied just over two minutes (if you are at the eclipse limits or near the end of the track you will have to reduce the frequency as appropriate, as totality is much shorter). Whatever you do, once you reach 4s exposure then pause and drink in the spectacle. You can always resume a little while later (if you are not too awe-struck). If you can, set the exposure back to 1/2000s, move the eclipse back to the centre of the viewfinder and wait for the final diamond ring. When it occurs, keep pressing the shutter as the diamond ring could last four to five seconds.

If you want to ignore this advice, simply reverse the sequence as totality ends to capture prominences on the other limb of the Sun and the final Baily's Beads and diamond ring. Once totality is finished, do not forget to replace the filter.

At sea, eclipse photography is constrained by the movement of the ship. It is unlikely that you could use a focal length of more than 500mm because of this. You must also contend with vibration from the ship's engines, wind across the deck, and other passengers' footsteps. On a ship it is best to use the ISO 1600 in order to be able to use the fastest shutter speeds possible. Notice the range of motion of the ship and attempt to snap the picture when it reaches one extreme.

For more specialist information see *Further Information*, page 48. Specialist telescope and camera shops will sell adaptors for connecting cameras to telescopes. Do not use the eyepiece filter that is furnished with some small telescopes – it is not safe.

HOW TO USE A CAMCORDER (OR DSLR WITH VIDEO FACILITY) TO CAPTURE THE ECLIPSE These can be used in two modes:

- To take the eclipse itself using the maximum zoom setting. Make a filter as described, mount the camera on a tripod and set the camera to automatic. Remove the filter during totality (or slightly before) and continue. If the camcorder has a manual mode, try increasing the exposure during totality to bring out the fine structure of the corona. Make sure any built-in illuminating light is disabled. Practise all this at home on the un-eclipsed Sun many weeks beforehand.
- To film the observing site and the people around, including their reaction to totality, put the camera on a tripod and set it to automatic with the zoom at its widest setting. Rotate the camera before and during totality. Make sure any built-in illuminating light is disabled. If the camera has a manual mode fix the exposure about ten minutes before totality and do not vary it. This way you will show the increasing level of darkness as the eclipse progresses.

7

Conclusion

Hopefully you may be reading this before you have seen these eclipses, in which case the conclusion for you has not yet happened, and what a conclusion it will be! Why not email me with details of your experience of totality (e *sheridan@eclipse.im*), how it affected you emotionally, and the reactions of others around you. Or take a look at the league table of umbraphiles at www.clock-tower.com/total.htm, and log the time you spend under the Moon's shadow. Marvel at the fact that some people have travelled to 29 total solar eclipses and only been clouded out once!

If you are reading this after the eclipse, you should now realise what all the fuss was about, and why people travel to the ends of the Earth to be submerged in totality, and why only totality will do. Once you have joined the club and seen more than one total eclipse you will almost certainly be hooked and want to see more.

Total eclipses are infrequent enough to allow time to save, and they offer the opportunity of travel to places that you might otherwise not consider. Take a friend who hasn't witnessed totality, especially as there is always so much else to see at your chosen destination.

If you've never seen an eclipse from a cruise ship, then do not rule it out. Long exposure photography is more difficult, but the camaraderie makes up for this. As almost all eclipses have large parts of their track over the oceans, you often have no alternative than to join a cruise (see opposite for examples).

The March 2016 eclipse will pass over Bangka Island, Philippines (OK/DT)

The total eclipse map on the back of this booklet is a good overview of the eclipse tracks until 2035, and you can use the information below as your holiday planner. I wish you success and clear skies in your eclipse chasing, and I'm sure you will enjoy meeting like-minded friends in the most unlikely places.

MARCH 2015 Using a bit of artistic licence, this is the next British eclipse. It just touches tiny Rockall Island and then tracks northward to one of the cloudiest places on Earth – the Faroe Islands. On it travels to Svalbard Island and ends exactly at the North Pole, which in March is not an easy destination to reach. A cruise or a flight is the only realistic way to witness this, and I don't mind betting that very few people actually see totality.

MARCH 2016 This is another eclipse I'm waiting for because it crosses Bangka Island where I saw my first total eclipse in 1988. It starts by crossing the Mentawai Islands in west Indonesia, then moves on to Sumatra, Bangka Island, Kalimantan, Sulawesi, and finally the Caroline Islands in the Pacific. It passes over a number of fantastic places to visit.

AUGUST 2017 Very unusually, mainland USA has been waiting for 36 years for totality, and this eclipse does it proud. It crosses California, Oregon, Idaho, Wyoming, Nebraska, Iowa, Missouri, Illinois, Kentucky, Tennessee, North Carolina, Georgia and South Carolina. Interestingly, the USA only has to wait seven years for the next one in 2024, which crosses the 2017 track in the Kentucky area. So if you want to be one of the privileged few to have seen two total eclipses from the same location, this is your chance.

JULY 2019 Another very difficult eclipse to view, this one passes mainly over the Pacific Ocean and Oeno Atoll near Pitcairn. It hits land in southern Chile, moving into Argentina. In July, however, these locations will be decidedly cool. Again, a cruise ship might be an option here.

8

Further Information

BOOKS

British Astronomical Association *Introduction to DSLR Astrophotography*. 2011. A splendid publication with all sorts of tips and ideas on how to photograph astronomical events.

Espenak, Fred and Anderson, Jay. Definitive scientific guides to any single upcoming eclipse are produced in the form of a bulletin, put together for the American space agency, NASA. They include predictions, tables, maps and weather prospects. NASA's eclipse website is http://eclipse.gsfc.nasa.gov/eclipse.html.

Folley, Tom and Zaczek, Ian *The Book of the Sun* Courage Books. To understand more of the folklore and mythology surrounding the Sun.

Guillermier, Pierre et al *Total Eclipses* Springer, 1999. For a book heavily weighted towards the science side, this has a less personal tone and is for those who really want to get into the history of the scientific study of eclipses.

Hoskin, Michael, ed *The Cambridge Illustrated History of Astronomy* Cambridge University Press, 1997. If humanity's past responses to eclipses has whetted your appetite for the history of astronomy try this book.

Littman, Mark, Willcox, Ken and Espenak, Fred *Totality: Eclipses of the Sun*, Oxford University Press, 1999. A well-written exposition of the science, culture and history of eclipses, set alight by the authors' enthusiasm.

McEvoy, J P *Eclipse: The Science and History of Nature's Most Spectacular Phenomenon* Fourth Estate, 1999. The most sugared scientific pill I have found, it is strong on the astronomical and cultural history of eclipses and easy to read.

Maddox, Robert *1927: A British Eclipse*. An intriguing and amusing account of the response of a nation to a total solar eclipse – from scientific preparations to day-trips. Includes some moving quotes from eyewitnesses, as well as many disaster stories, most of them due to the fickle weather. It can be ordered from the British Astronomical Association, Burlington House, Piccadilly, London W1V 9AG for £3.50, postage free in Britain and Europe.

Pasachoff, Jay and Menzel, Donald *A Field Guide to the Stars and Planets* Houghton Mifflin, 1992 Find out more about the heavens fleetingly revealed during the eclipse.

Stephenson, Richard F *Historical Eclipses and Earth's Rotation* Cambridge University Press, 1997. Full of interesting quotations about eclipses as well as following – and solving – an intriguing scientific mystery using the eclipse calculations of the Chaldeans.

Williams, Sheridan *UK Solar Eclipses from Year 1 to 3,000* Clock Tower Press, 1996. Contains an anthology of 3,000 years of solar eclipses – an intriguing catalogue of the stories attached to Britain's eclipses including those that have started wars, accompanied new kings and symbolised death. Also included are all the eclipse facts you could possibly wish for, a detailed look at eclipse mechanics and astronomical software, and discussion of the practicalities of viewing eclipses. There are 21 beautiful eclipse photographs. The book can be obtained through bookshop or direct from the publisher. To order via the internet go to www.clock-tower.com/eclipse.htm. You can also send a sterling cheque drawn on a UK bank and payable to Sheridan Williams, for £5.95 plus p&p: £1 in UK, £2 Europe/Eire, £3 elsewhere to: Clock Tower Press, The Clock Tower, Stockgrove Park, Leighton Buzzard, Bedfordshire LU7 0BG, England.

DVDS

The Society for Popular Astronomy has a 45-minute highly instructive DVD available from www.popastro.com or write to SPA, 36 Fairway, Keyworth, Nottingham NG12 5DU. These are available to members, but at £15 to join, there is a huge amount to gain and learn. **BAA 2006, 2008, 2009 and 2010 Eclipse DVDs** are available from the British Astronomical Association, which is the voice of amateur astronomy and is also worth a subscription to obtain eclipse material. Further details can be found at www.britastro.org.

INTERNET

The internet is packed with eclipse information compiled by dedicated chasers, who are a technologically literate lot.

http://tinyurl.com/6d57y2o Introduction to digital astrophotography from the British Astronomical Association.
www.astronomy-uk.co.uk Secondhand astronomy equipment.
http://sunearth.gsfc.nasa.gov/eclipse/toplink.html This site lies at the heart of all this internet activity. It includes a formidable amount of information produced by scientists who calculate eclipse paths, plot them on maps and deposit them online along with weather forecasts and even analyses of the likelihood of forest fires at your chosen viewpoint. This site will also provide you with links to all the other sites you could possibly want.
http://umbra.nascom.nasa.gov/eclipse/ For the astronomically minded, the official NASA solar eclipse bulletin site.
www.williams.edu/Astronomy/IAU_eclipses/ For the uses of eclipses to science.
www.earthview.com/ For tutorials.
www.mreclipse.com/ If it is fascinating facts you are after, such as how to say total solar eclipse in over 80 different languages, and view eclipse stamps over the last 50 years this is your site. It is run by Fred Espenak.
www.clock-tower.com/eclipse.htm For British eclipse links.
http://eclipse.im For links to all eclipse sites worldwide.
www.eclipsechaser.com Most of the eclipse tour companies have websites; this is a central site with links to many of them and also includes testimonies and photographs of past eclipses.
http://en.wikipedia.org/wiki/Transit_of_Venus For the transits of Venus.
www.willbell.com/math/mc13.htm Transits by Jean Meeus – an extremely good site which also includes transits of Mercury and also transits of other planets from other planets, such as a transit of Uranus as seen from Neptune.
http://astro.ukho.gov.uk/nao/transit/V_2012/ HMSO transit predictions.

http://xjubier.free.fr/en/index_en.html Xavier Jubier's eclipse site with Google Earth maps and links.

WEATHER WEBSITES
Forecasts and sources of satellite imagery
https://afweather.afwa.af.mil/weather/satellite.html USAF weather site with satellite imagery over several sectors of the globe.

www.sat.dundee.ac.uk Dundee University Satellite Receiving Station. One of the most comprehensive sites, Dundee has images from geostationary satellites around the globe at high resolution. Images are not rectified and have only minimal country outlines and latitude markings, so that finding a particular site requires considerable care. Registration (free) is required.

www.weatherzone.com.au/satellite.jsp Real-time satellite images and animations of Australia from WeatherZone.

www.bom.gov.au/weather/qld/ Queensland forecasts from the Bureau of Meteorology in Australia.

www.bom.gov.au/weather/nt/ Northern Territory forecasts from the Bureau of Meteorology.

http://realtime.bsch.au.com/currentwx.html High-resolution satellite images from the Brisbane Storm Chasers website.

www.eclipser.ca Climatology and maps for the eclipse chaser by meteorologist Jay Anderson.

MAPS

Maps of countries are available from many sources. Two of the best are:

Stanford's 12–14 Long Acre, London WC2E 9LP; www.stanfords.co.uk. Tactical Pilotage Charts (1:500,000)

Maps Worldwide Datum Hse, Lancaster Rd, Melksham, Wiltshire SN12 6TL; 01225 707004; www.mapsworldwide.com

http://xjubier.free.fr/en/site_pages/SolarEclipsesGoogleMaps.html For interactive web-based maps, Xavier Jubier has plotted future eclipse tracks on Google Earth.

www.eclipse-maps.com/Eclipse-Maps/Welcome.html A site showing ancient and modern eclipse maps.

WIN A FREE BRADT GUIDE
READER QUESTIONNAIRE

Send in your completed questionnaire and enter our monthly draw for the chance to win a Bradt guide of your choice.

To take up our special reader offer of 40% off, please visit our website at www.bradtguides.com/freeguide or answer the questions below and return to us with the order form overleaf.

(Forms may be posted or faxed to us.)

Have you used any other Bradt guides? If so, which titles? ....
....

What other publishers' travel guides do you use regularly? ....
....

Where did you buy this guidebook? ....

What was the main purpose of your eclipse trip? eg: holiday/business ....
....

How long did you travel for? (circle one)

| weekend/long weekend | 1–2 weeks | 3–4 weeks | 4 weeks plus |

Which countries did you visit in connection with this trip? ....
....

Did you travel with a tour operator?' If so, which one? ....
....

What other destinations would you like to see covered by a Bradt guide?
....

If you could make one improvement to this guide, what would it be?
....

Age (circle relevant category)   16–25   26–45   46–60   60+

Male/Female (delete as appropriate)

Home country ....

Please send us any comments about this guide (or others on our list).
....
....
....

Bradt Travel Guides
IDC House, The Vale, Chalfont St Peter, Bucks SL9 9RZ, UK
↘ +44 (0)1753 893444 **f** +44 (0)1753 892333
e info@bradtguides.com
www.bradtguides.com

TAKE 40% OFF YOUR NEXT BRADT GUIDE!
Order Form

To take advantage of this special offer visit www.bradtguides.com/freeguide and enter our monthly giveaway, or fill in the order form below, complete the questionnaire overleaf and send it to Bradt Travel Guides by post or fax.

Please send me one copy of the following guide at 40% off the UK retail price

No	Title	Retail price	40% price
1	....	....	....

Please send the following additional guides at full UK retail price

No	Title	Retail price	Total
...	....	....	....
...	....	....	....
...	....	....	....

Sub total ....
Post & packing ....
(Free shipping UK, £1 per book Europe, £3 per book rest of world)
Total ....

Name ....
Address ....
Tel .... Email ....

☐ I enclose a cheque for £........ made payable to Bradt Travel Guides Ltd

☐ I would like to pay by credit card. Number: ....

   Expiry date: .../....... 3-digit security code (on reverse of card) .....

   Issue no (debit cards only) .....

☐ Please sign me up to Bradt's monthly enewsletter, Bradtpackers' News.

☐ I would be happy for you to use my name and comments in Bradt marketing material.

Send your order on this form, with the completed questionnaire, to:

Bradt Travel Guides
IDC House, The Vale, Chalfont St Peter, Bucks SL9 9RZ, UK
☎ +44 (0)1753 893444   f +44 (0)1753 892333
e info@bradtguides.com   www.bradtguides.com